BEHAVIORAL OBJECTIVES AND EVALUATIONAL MEASURES

Science and Mathematics

BEHAVIORAL OBJECTIVES AND EVALUATIONAL MEASURES

Science and Mathematics

Robert B. Sund
UNIVERSITY OF NORTHERN COLORADO

Anthony J. Picard
UNIVERSITY OF HAWAII

CHARLES E. MERRILL PUBLISHING COMPANY
A Bell & Howell Company
Columbus, Ohio

International Standard Book Number: 0-675-09761-4

Library of Congress Catalog Card Number: 74-161434

1 2 3 4 5 6 7 8 9 10/76 75 74 73 72

Printed in the United States of America

PREFACE

Educators are becoming increasingly aware of the role of behavioral objectives in the educational process. Recent national conventions of the American Educational Research Association, the National Science Teachers Association, and the National Council of Teachers of Mathematics devoted several sections to papers on the subject. The advent of individualized instruction and computerized educational programs have stimulated the creation of behavioral objectives for complete educational packages. State educational committees, school districts, administrators, and teachers are devoting thousands of hours to developing them. Some federal granting agencies require the specification of performance criteria for the projects they fund. Banks of objectives and related test items are being constructed in the United States, Canada, Great Britain, and other countries. Articles on behavioral objectives permeate educational journals.

This book endeavors to familiarize the reader with the dialogue presently underway concerning the relevance and value of behavioral objectives. Included are (1) an outline of the main categories of objectives, i.e., cognitive, affective, psychomotor, and performative; (2) criteria for their selection, simple practical suggestions on how to write them, and illustrative examples of science and mathematics objectives from elementary to college levels; (3) examples of how to evaluate pupil achievement relative to the objectives, including numerous tests and instruments for analyzing an examination, a course, and teacher behaviors.

The authors have attempted to write a clear, practical book giving many helpful suggestions, particularly in the use of self-evaluation inventories and evaluative techniques for the affective domain. Many of the examples used in this book have been written by students preparing to teach or by experienced teachers. These examples are included to indicate that extensive training is not required to write behavioral objectives. However, as the reader produces objectives of his own, his competence in this skill will increase and he will become more aware of the intricacies of excellent teaching. Because objectives and tests are included from many individuals, there is considerable variation in their sophistication. The reader should be cognizant of this variation, consider how the objectives might be improved, and adapt them to his situation.

Learning to write behavioral objectives is an important technique for improving instructional competence. This is especially true with the writing of affective domain objectives. Teachers creating these become more aware of their influence in the instructional process and are, therefore, more likely to modify their teaching behavior to more affective ends.

We would like to acknowledge the assistance of many national curriculum projects, school districts, individual teachers, and our own students for their help and criticisms in the writing of this book. Without their help and encouragement this publication would not have been possible.

R.B.S.
A.J.P.

CONTENTS

CHAPTER ONE

THE PLACE OF BEHAVIORAL OBJECTIVES

What Are Behavioral Objectives?

Behavioral objectives are receiving increasing attention by educators. Teachers and administrators are being asked to justify the entire school curriculum in behavioral statements. Reference to behavioral objectives is found in most recent issues of professional education journals. But, even as its usage becomes more popular, a precise meaning of "behavioral objective" is, in many cases, lacking.

Some educators use the expression behavioral objective interchangeably with the words "aim" or "goal." Others restrict its usage to situations in which some clearly defined behavior by the learner is observable. Although the expression "behavioral objective" evokes a general meaning, a precise definition is necessary in order to distinguish behavioral objectives from these other terms.

What Is the Difference Between An Aim or Goal and A Behavioral Objective?

Aims or Goals

These are broad general statements, sometimes vague in meaning, which generally shape the character of an educational program.

Examples of Goals

1. To develop proficiency in thinking through problems and interpreting data — graphs, charts, and tables — especially as they pertain to home, business, and community.
2. To apply the scientific mode of inquiry to problems relating to the physical world.

Purpose of Goals

1. Goals are necessary to establish an overall climate of the classroom.
2. Goals may be used as a guide for writing specific behavioral objectives.

Behavioral Objectives

These are objectives written in behavioral terms. Behavior may be defined as visible activity. Behavioral objectives state how a person is to act, think, or feel.

Examples of Objectives

1. When presented with a graph taken from a newspaper, the child is able to identify, using ordered pairs, three points which are on the graph.
2. The child is able to identify the stages of mitosis and to describe in his own words the changes which occur in the cell at each stage.

Statements which begin "to teach" or "to introduce" are not objectives in terms of the definition above. These statements refer to the teacher's behavior rather than to the students'. Curriculum guides which merely list the course content are also not statements of objectives because they omit the behaviors the successful learner should exhibit when evaluated in terms of that content. An objective is not:

1. The teacher's behavior
2. A list of the course content
3. A life value (6, p. 30)

The language used to state an objective must be as precise as possible. Any ambiguities in wording will obscure what the writer expects the learner to do when the objective has been achieved. "An objective is meaningful to the extent it communicates an instructional intent to the reader and does so to the degree it describes or defines the terminal behavior expected of the learner" (9, p. 43).

Words Used to State Objectives

Poor	Better
know	recall, identify, define, distinguish, describe,
understand	demonstrate, compare, contrast, classify,
apply	complete, select, assign

An objective differs from an aim or goal by stating explicitly the kind of performance a student must demonstrate as evidence that learning has taken place. In effect, an objective may be thought of as an aim having been translated into clearly identifiable components (5, p. 83) or as the intermediate steps in the achievement of broader, more general statements of goals (7, p. 12). An objective usually states *what* the learner is to be like, while an aim states *why* he should behave in this manner (15, p. 83).

What Are the Reasons for Increased Interest in Behavioral Objectives?

The main reasons for the increased interest in behavioral objectives may be summarized as follows:

1. The influence of behavioristic psychology on instruction.
2. The need for behavioral objectives in programmed units and computer assisted instruction. Although the behavioristic theory of learning dates from the 1930's, its current popularity may be traced to the greater emphasis placed on teaching machines and programmed learning in today's educational situations. Because programmed units require specific step-by-step prescriptions of learner responses, they have recreated an interest in behavioral objectives. Psychologists and educators usually associated with this movement are James Popham, Ralph Taylor, Robert Gagné, Robert Glaser, and Robert Mager.
3. The lack of well-written objectives in contemporary textbooks which would answer questions such as:

   a. Is the book mainly a self-contained instructional device?
   b. How are auxiliary materials to be supplemented with the text?
   c. How is the student who has learned the materials identified?
4. The need to identify the depth of comprehension of the successful student.
5. The lack of emphasis on higher thought processes in traditional curricula.
6. The need to identify the important substance of the course.
7. The rising concern over the lack of emphasis of affective and psychomotor aspects of instruction.
8. The increased involvement of teachers in curriculum reform. Every teacher formulates objectives. Even teachers not writing down their objectives have planned, consciously or subsconsciously, the direction and purpose of their lessons. But, objectives that are only mental images are too vague for the precise task of curriculum construction. The statement of objectives in terms of learner behaviors requires the teacher to translate ideas about how the student should behave into clearly identifiable actions.
9. The need to make curriculum guides more useful. Many state and national organizations have recognized the inadequacy of curriculum guides listing simply the content to be covered at a particular grade level. The inclusion of behavioral objectives into these guides provides the teacher with more direction for classroom activities and evaluation. Colorado, California, Hawaii, and Wisconsin are a few states where K–6 curriculum guides involving statements in behavioral terms have been developed for mathematics or science.

Behavioral Objectives and the Instructional Process

The idea of a goal toward which the educational experience should be directed is as old as education itself. The attainment of "the good life" or "truth" are two of the themes man has used to plan, organize, and direct his learning environment. Objectives are only one of the diverse components of the instructional process. At what level of planning do they enter this process?

The Instructional Process

A MODEL OF THE INSTRUCTIONAL PROCESS

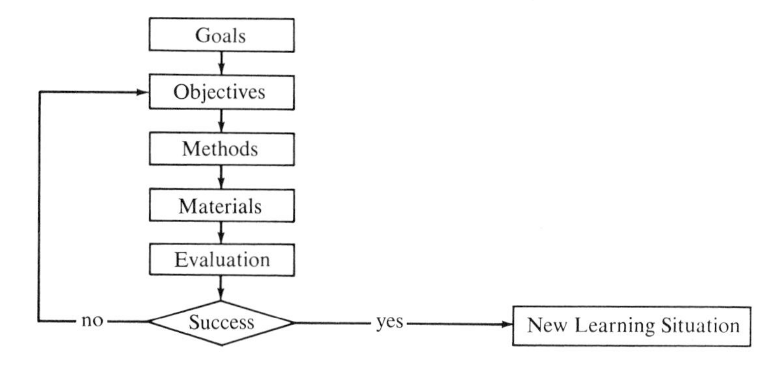

The instructional process may be viewed as a flow chart in which each content unit represents a loop. The goals determine the general atmosphere. The objectives, specifying both content and behavior, indicate what the pupil should be able to do at the end of the unit. The objectives suggest certain methods of presentation, and the activities require certain materials and teaching aids. The evaluation of pupils is determined by the objectives established at the beginning of the unit.

After evaluation, the program branches. If the evaluation shows the pupil has successfully achieved the objectives, he is ready to go on to a new learning situation. If he is unsuccessful, the teacher must re-evaluate the objectives to determine their appropriateness. An alternate teaching strategy and new experiences must be designed for the learner. The evaluation must be examined to see if the behaviors required in the measuring instrument are the same ones required in the objectives. After the necessary modifications are made, the pupil proceeds through the loop a second time. The process of evaluation and modification is repeated until the learner is successful.

The advantages of using behavioral objectives in the instructional process is that they describe the behaviors the student should demonstrate at the end of the process. Since evaluation is based on whether or not the student can perform these behaviors, the exact areas of failure can be located and communicated to the student. Furthermore, the failure to attain the objectives may be used as a diagnostic instrument for modifying the instructional design. The modification process becomes less haphazard and leads to specific alterations of the learning situation.

Another Model of the Instructional Process

Goals Objectives	What is it you want the student to do?
Methods Materials	What experiences, situations, and teaching aids elicit these behaviors?
Evaluation	Have the desired behaviors been achieved?

In addition to providing a more valid basis for student evaluation and for revising learning activities, behaviorally stated objectives benefit the learner in another way. Students are continually accused of studying for the test rather than for the sake of knowledge. Mager tells of an Army training course in which all the students did poorly every third test. An analysis of the situation showed the course had been arranged into three week units with a new instructor for each unit and a test each week. Since the objectives of the course were not known to the students, they had to wait until the first test of each unit to find out what was expected of them (9, p. 3). This is not an isolated case. Students are quite capable of determining the objectives of a course from the examinations used to evaluate them (11, p. 81). An instructor may attempt to incorporate many objectives into the learning experience, but the ones he uses to construct his tests determine what the students study.

One way of avoiding this situation is to provide the learner with a copy of the objectives at the beginning of the instructional process. If clearly stated in behavioral terms, these objectives indicate exactly what the learner must do in order to show he has achieved them. It follows that the measuring instruments must call for exactly the same behaviors stated in the objectives. In this way, the learner knows at the beginning of the learning situation how he must perform in order to be successful. "Studying for the test," in this case, amounts to being able to perform the behaviors expected of him.

The notion that a learner achieves more if he knows the objectives of the course has been an educational maxim for some time. Behavioral objectives offer a unique opportunity to test this hypothesis. There is currently at least one project, CAM (Comprehensive Achievement Monitoring), comparing student achievement with familarity of the objectives. An important by-product of this approach appears to be a reduction in learner anxiety (12, p. 208). Once the student discovers the test measures exactly what the objectives call for, there is a tendency to concentrate on the list of objectives. The actual test becomes less important than being able to perform according to the objectives. This project will be discussed more fully in chapter six.

Criticisms of Behavioral Objectives

The following list summarizes some of the main criticisms leveled against behavioral objectives by various critics and gives some opposing views in answer to these:

1. The time spent writing objectives could be better spent in planning instruction.
2. All behavioral objectives may not be capable of being defined (12, p. 233).
3. Critics believe some objectives such as creativity, problem solving, etc., cannot be broken down into component behaviors and still have meaning.
4. The formation of intermediate behaviors is no guarantee that these can be synthesized to form terminal general behaviors. While much research needs to be done in this area, beginning efforts have shown there is a higher incidence of terminal behavior in learners having acquired the intermediate behavior than in those who have not (15). The *Science — A Process Approach* Curriculum of the American Association for the Advancement of Science has shown with careful investigation that many of the most elusive behaviors can be analyzed into component behaviors. In addition, certain behaviors may be given operational or procedural definitions. This type of definition is an agreement that in a given situation, certain learner activities will be accepted as evidence that the behavior has been achieved. For example, problem solving might be defined operationally as the ability to produce a reasonably accurate solution when presented with an unfamiliar problem.
5. Certain objectives cannot be reliably measured. The answer to this criticism lies in the acceptance of operational definitions and in greater diversity of measurement. Multiple types of evaluation should be used. It is no longer sufficient to rely solely on "paper and pencil" tests. Interview, self-evaluation, and observation techniques have achieved a degree of validity and reliability warranting their use to assess learner achievement. The topic of evaluation, including observation, self-evaluation, and interview techniques, will be discussed in chapters eight and nine.
6. Since it can be argued that observable behavior does not necessarily correspond to internal mental activity, there is a danger of making invalid inferences about the learner's mental achievement if too much emphasis is placed on observable behaviors.

However, this danger exists in any evaluation and is not peculiar to a behavioral approach.

7. Behavioral objectives may obscure goals. David Ausbel, a leading proponent of this view says:

> . . . behavioral terminology more often obscures than clarifies educational goals. *The Taxonomy of Educational Objectives* (Bloom, 1956, Krathwohl, et. al., 1964), for example, categorizes educational goals in great behavioral detail. But since such terms as "memory," "application," "understanding," "transfer," "meaning," "cognitive," and "affective" have different meanings for practically every practicing psychologist, the classification of curriculum objectives along such lines merely results in considerable pseudo-agreement among psychologists and curriculum workers without ever defining what the actual objectives in question are. Every one is happy because of the fine degree of "scientific" precision achieved in defining goals . . . but nobody seems to care whether this achievement is psychologically or educationally meaningful (1).

The use of *The Taxonomy of Educational Objectives* as a guide for creating objectives is clearly not the use intended by Bloom and his associates. They point out that *The Taxonomy* is to be used to classify objectives and such classifications may vary according to the student's background and experiences. Numerous experiments conducted over the past years have generally substantiated the simple to complex and concrete to abstract hierarchical foundation of *The Taxonomy* (4).

Perhaps the most valid criticism of the behavioral approach is that it narrows the teacher's view of learning. Education is reduced to industrial engineering, with quality control and systems analysis the bases for decision making. This concentration on small bits of learning lends itself to an undue emphasis on easily defined, but relatively trivial, objectives.

It is a healthy sign that not all educators hold a behavioristic view. The renewed emphasis on behavioral objectives has forced behaviorists and nonbehaviorists to reexamine their positions. This kind of activity is necessary for the growth of the teaching profession.

Classification of Behavioral Objectives

Numerous efforts have been made to formulate a classification scheme, a taxonomy, for educational objectives. The first efforts essentially divided objectives into two groups — general and specific.

General objectives are those of concern throughout the entire period of instruction. In science and mathematics, they include such things as problem solving, critical thinking, and creativity.

Specific objectives are usually limited to the concepts and principles of the subject matter. They might include such things as number, numeration systems, and geometric facts in mathematics and concepts such as the cell, lever, simple machines, lenses, and magnets in science. Principles such as: when heated, metals expand; the basis of life is the cell; the order of summation of two numbers doesn't affect the sum; and the heating of a chemical reaction increases the speed of the reaction; fall into the same category.

Teachers have accepted general objectives as desirable aims for years. However, the vagueness of these terms gave instructors little insight into implementing them. Concerned about the haphazard ways objectives were selected and stated, a group of psychologists met in 1948 to establish "a common language for classifying human behavioral characteristics" (2, p. 3). After a brief investigation, this group soon realized objectives could be classified into three categories or domains, which they labeled cognitive, effective, and psychomotor.

The group constructed a taxonomy based on the assumption that objectives could be arranged along a continuum from concrete to abstract and from simple to complex. It also assumed any level of the taxonomy would necessarily include the objectives of all the levels lower than that one. An abbreviated form of this taxonomy is given below:

1. Cognitive domain — objectives dealing with the recall or recognition of knowledge and the development of intellectual abilities and skills.

   a. Knowledge (Recall)
   b. Comprehension
   c. Application
   d. Analysis
   e. Synthesis
   f. Evaluation

2. Affective domain — objectives dealing with changes in interest, attitudes, and values, and the development of appreciations and adequate adjustments.

   a. Receiving
   b. Responding
   c. Valuing
   d. Organization—conceptualization and organizing a value system
   e. Characterization by a Value or a Value Complex

3. Manipulative or Psychomotor Domain—objectives dealing with muscular or motor skills such as the manipulation of apparatus.

The authors of *The Taxonomy* divided each level into sub-levels. For example, cognitive level 2, comprehension, is divided into translation, interpretation, and extrapolation. They pointed out there is considerable overlapping of adjacent levels and so the location of an objective may vary.

Some psychologists argue that although this taxonomy contributes considerably to the classification of objectives, the three domains should not be thought of as separate and unique in operation. For example, when a student is learning some scientific or mathematical concept or principle (cognitive objective), he simultaneosuly is developing certain attitudes and values toward the concept he is learning (affective objective). Butts, in an analysis of research, states:

> Although cognitive behavior is more clearly identified as a faculty of knowing as distinguished from the influence of feeling, there is considerable research indicating cognitive and affective behavior cannot be completely separated (Bloom and Broder, 1950, Johnson, 1955, Russell, 1956, Wertheimer, 1945) . . . A review of the literature suggests that factors of cognitive behavior may include problem solving, perceptual closure, verbal fluency, ideational fluency, and tested intelligence. Affective behavior, though difficult to separate from cognitive behavior, may include attitude and selectivity of perception. The latter may be related to attending behavior. Both cognitive and effective behavior appear to be related to environment (3, pp. 2, 18).

Krathwohl, Bloom, and Masia, in discussing the relationship between cognitive and affective domains, say:

> We recognize that human behavior can rarely be neatly compartmentalized in terms of cognition and affect. It is easier to divide educational objectives and intended behaviors into these two domains. However, even the separation of objectives into these two groups is somewhat artificial in that no teacher or curriculum worker really intends one entirely without the other (8, p. 85).

Because cognitive, affective, and psychomotor objectives often cannot be isolated in actual human operations, Louise L. Tyler and Laura J.

Okumu have suggested that the term "performative objective" be established (13). Performance objectives involve the acting out of a set of behaviors. Therefore, they are more encompassing and may include cognitive, affective, and psychomotor aspects. Examples of this type of objective are listed below.

The student should:

a. Act as though he seeks new information about his physical environment, e.g., weather, earth, and sky, and report his findings to the class.
b. Attempt to design an experiment when given a problem containing some recognizable elements.

These objectives could be classified according to Bloom under the category of a Value Complex of the Affective Domain. Analysis of the objectives, however, will indicate they involve both cognitive behaviors (knowing something about the physical sciences, how to seek information and how to design an experiment) and affective behaviors (seeking information and wanting to design an experiment). The student successfully demonstrating these behaviors would be acting like a scientist. The awareness of these performative objectives helps teachers to construct and look for more inclusive behaviors. By being aware of them, teachers tend to understand better the interrelations of cognitive, affective, and psychomotor aspects of the learning process.

Objectives of this Book

After completing the study of this book, the reader should be able to:

1. State a definition of aim, behavior, and behavioral objective.
2. Distinguish between an aim and a behavioral objective.
3. State several criticisms and advantages of behavioral objectives.
4. Reproduce a behavioral model for the instructional process.
5. State several behaviors common to science and mathematics.
6. List a set of criteria for selecting objectives in science and mathematics appropriate to a given situation.
7. Classify objectives of science and mathematics as cognitive, affective, or psychomotor (according to *The Taxonomy of Educational Objectives*).

8. List a set of criteria for stating behavioral objectives.
9. Write behavioral objectives in science and/or mathematics.
10. Outline the process of constructing a curriculum based on behavioral objectives.
11. Construct instruments to measure behaviors in science and mathematics at all cognitive levels.
12. Determine the distribution of items of a given test among the categories of Bloom's *Taxonomy of Educational Objectives* (Handbook I) using a content behavior grid.
13. Construct tests using a content behavior grid.
14. Evaluate science and mathematics teaching performance according to specific criteria, using self-analysis and interaction analysis techniques.
15. Evaluate science and mathematics curricula.

Summary

A behavioral objective is a statement of a desired change in student behavior as a result of participating in some educational experience. Behavioral objectives differ from more general statements called aims or goals by specifying, in terms of observable student behavior, what the successful learner must be able to do after instruction has taken place.

The statement of behavioral objectives in terms of learner behaviors originated with the behaviorist school of psychology. Recently, a renewal of interest in behavioral objectives has been created by programmed materials and by the demand for teachers to participate in curriculum development.

Behaviorally stated objectives provide a basis for a model of the instructional process. They help reveal insights into alternate teaching strategies and learner activities, suggest certain appropriate learning aids, provide for valid evaluation of student progress, and serve as a starting point for curriculum construction and modification.

Behavioral objectives have been criticized for narrowing the teacher's view of the educational experience. Too much emphasis is placed on easy to state, but relatively trivial behaviors. More complex behaviors cannot be broken down into readily achievable component behaviors. While some of these criticisms are valid, progress is being made toward the solution of the problems from which they arise.

Bibliography

1. Ausubel, D. "Psychological Aspects of Curriculum Evaluation." Paper presented to the National Seminar for Vocational Education Research, University of Illinois, May 18, 1966.

2. Bloom, et al. *Taxonomy of Educational Objectives, the Classification of Educational Goals, Handbook I: Cognitive Domain*. New York: David McKay Co., Inc., 1956.

3. Butts, David. "Science Inservice Project Research Report No. 4." Unpublished paper, Research and Development Center for Teacher Education, Austin, Texas: University of Texas, 1968, p. 2.

4. Cox, R. and Unks, N. "A Selected and Annotated Bibliography of Studies Concerning *The Taxonomy of Educational Objectives: Cognitive Domain*." Working Paper #13. Learning Research and Development Center, University of Pittsburgh, 1967, pp. 2-33.

5. Eisner, E. "Instructional and Expressive Educational Objectives: Their Formulation and Use in Curriculum." Monograph, American Educational Research Association.

6. Furst, E. *Constructing Evaluation Instruments*. New York: Longmans, Green and Company, 1957.

7. Johnson, D. and Rising, G. *Guidelines for Teaching Mathematics*. Belmont, California: Wadsworth Publishing Company, Inc., 1967.

8. Krathwohl, D., Bloom, B., and Masia, B. *Taxonomy of Educational Objectives, the Classification of Educational Goals, Handbook II, Affective Domain*, New York: David McKay Co., Inc., 1964.

9. Mager, R. *Preparing Instructional Objectives*. Copyright 1962 by Fearon Publishers/Leor Siegler, Inc., Education Division. All rights reserved. Reprinted by permission, pp. 43, 45.

10. Ojemann, R. "Should Educational Objectives be Stated in Behavioral Terms? Part I." *The Elementary School Journal*, February 1969, pp. 229-35.

11. Picard, A. "An Analysis of the Objectives of a First Year Calculus Sequence, a Test for the Achievement of These Objectives, and an Analysis of Results." Unpublished doctoral dissertation, Ohio State University, 1967.

12. School and Society, *Evaluation: Instant Measurement*, April 1, 1967, pp. 208-09.

13. Tyler, L. and Okumu, L. "A Beginning Step: A System for Analyzing Courses in Teacher Education. *Journal of Teacher Education,* December 1965.

14. Walbesser, H. "Some Properties of Interesting Space-Behavioral Objectives, Learning Hierarchies, Mathematical Curricula." Paper presented at the National Council of Teachers of Mathematics forty-sixth Annual Meeting, Philadelphia, Pennsylvania, 1968.

15. Wood, R. "Objectives in the Teaching of Mathematics." *Educational Research,* vol. X, no. 2, February 1968, pp. 83-98.

Chapter Two

GOALS OF SCIENCE AND MATHEMATICS

The Need for Selecting Goals before Writing Objectives

Before behavioral objectives can be written, the general goals of the curriculum should be outlined. This is advisable because in writing behavioral objectives the mechanics of construction often interfere with the consideration of their importance. For example, a writer may think it important for students to develop an enthusiasm for the course, but because he is unable to translate his view into behavioral terms easily, he rejects the goal. Defining goals first, therefore helps the instructor guard against the tendency to write mundane objectives.

The Goals of Mathematics

Identifying the goals of mathematics may appear a major task at first. Fortunately, however, there has been considerable work done in this area already. Donovan A. Johnson and Gerald F. Rising have outlined goals for mathematics, and their conclusions appear below:

> In selecting appropriate goals for mathematics instruction, we must take into account not only the needs of society, but also the mathematical needs of our students. Almost every committee or commission that has worked on revising the mathematics curriculum

has stated certain basic mathematical needs. The following list summarizes these needs:

1. The student needs to know how mathematics contributes to his understanding of natural phenomena.
2. He needs to understand how he can use mathematical methods to investigate, interpret, and make decisions in human affairs.
3. He needs to prepare for a vocation in which he utilizes mathematics as a producer and consumer of products, services, and arts.
4. He needs to understand how mathematics, as a science and as an art, contributes to our cultural heritage.
5. He needs to learn to communicate mathematical ideas correctly and clearly to others. Communication is a tool basic to all civilizations.

While it is not a specific mathematics objective, the teacher should not lose sight of the fact that the student needs to understand human relationships and to develop a personal value scale which respects the rights, needs and achievements of others (5, p. 11).

By analyzing selected "reports, recommendations, and standard examinations issued in the United States in recent decades" (4, p. 93), the Committee of International Study of Achivement in Mathematics identified the following ten important mathematical behaviors:

1. Ability to remember or recall definitions, notations, operations and concepts.
2. Ability to manipulate and compute rapidly and accurately.
3. Ability to interpret symbolic data.
4. Ability to put data into symbols.
5. Ability to follow proofs.
6. Ability to construct proofs.
7. Ability to apply concepts to mathematical problems.
8. Ability to apply concepts to non-mathematical problems.
9. Ability to analyze and determine the operations which may be applied.
10. Ability to invent mathematical generalizations (4, p. 93).

The Cambridge Conference on School Mathematics has had a great impact on mathematics education because of the boldness and scope of its recommendatoins. In its original report, published in 1963, the committee listed these goals for mathematics:

1. Lest there be any misunderstanding concerning our viewpoint, let it be stated that reasonable proficiency in arithmetic calculation

and algebraic manipulation is essential to the study of mathematics. However, the means of imparting such skill need not rest on methodical drill. We believe that entirely adequate technical practice can be woven into the acquisition of new concepts. But our belief goes farther. It is not merely that adequate practice can be given along with more mathematics; we believe that this is the only truly effective way to impart technical skills.

• • • • • •

2. Familiarity is our real objective. We hope to make each student in the early grades truly familiar with the structure of the real number system and the basic ideas of geometry, both synthetic and analytic.

• • • • • •

3. Mathematics is a growing subject, and all students should be made aware of this fact. This recommendation is not made merely because we feel that every educated person should know the fact, but also because the knowledge that there are unsolved problems and that they are gradually being solved puts mathematics in a new light, strips away some of its mystique, and serves to undermine the authoritarianism which has long dominated elementary teaching in the area.

• • • • • •

4. The building of confidence in one's own analytical power is another major goal of mathematics education.

• • • • • •

5. The function of language is to communicate. In mathematics its function is to communicate with extraordinary precision; it is inevitable therefore that mathematics requires some special terminology. Special terms are good or bad exactly according to their effectiveness in communication, and the same applies to special notations and symbols. This principle must not only guide textbook writers, it must be brought home to the student.

• • • • • •

6. To foster the proper attitude toward both pure and applied mathematics we recommend that each topic should be approached intuitively, indeed through as many different intuitive considerations as possible. In such a program, the student must be kept informed of where he stands. A curriculum which oscillates between logical rigor and guesswork can be confusing unless the student knows the level at all times. To present mathematics entirely in the rigorous deductive spirit not only precludes any possibility of applying mathematics, it is dishonest, even as a picture of contemporary pure mathematics. We hope that many problems can

be found (we know a few) that read, "Here is a situation — think about it — what can you say?"

• • • • • •

7. Another goal of our program is the inculcation of an understanding of what mathematics is (and what it is not). We need not here belabor the point that the man in the street has considerable misinformation on this point; suffice it to say that this misunderstanding frequently seems to take the paradoxical form of ascribing both too much and too little power to mathematics.

• • • • • •

8. While everyone should know about the wide range of topics suitable for mathematical analysis, it is almost equally important to understand the limitations of mathematics. Every application of mathematics depends on a model, and the value of a deduction is more an attribute of the model than it is of mathematics. We believe that students can be made aware of the distinction between the real world and its various mathematical models; in this we can look forward to cooperation from the sciences (2, pp. 7–12).

In a later conference, this group said:

> With respect to "Mathematics for mathematics," long-range goals were set forth during the 1963 Cambridge Conference on School Mathematics. Those goals can be accepted in the present report within the framework of the combined science-mathematics curriculum with two major qualifications. The first of these is that the mathematics should, especially during the early elementary school years, draw largely upon the science context both for its motivation and for the concrete situations that it models. This is not contrary to the 1963 report, for the earlier conferees could not assume the relevance, or even the existence, of the science content. Secondly, now that many new objectives have been added, the 1963 "goals" must for the time being be thought of as ideal objectives, to be attained only to the extent that the learning process can be accelerated by the integrated units and the pedagogical methods suggested in this report (3, pp. 19–20).

The Goals of Science

Science has been defined as both a body of knowledge and a set of processes or operations. The body of knowledge consists of the concepts and principles of science. The processes of science are often included under the term "problem solving."

In 1965 and later in 1967, Science Research Associates brought together panels to deliberate and describe science objectives. Educational Testing Service in 1968 held a similar conference, inviting leading scientists and educators to outline the science objectives so tasks could be defined for use in the National Assessment Tests. The statements of the objectives from these two groups were remarkably similar.

Raymond E. Thompson reports that the following four primary objectives evolved for use by the assessment project:

1. Know a reasonable amount about the fundamental facts and principles of science.
2. Possess to a reasonable degree the abilities and skills needed to engage in the process of science.
3. Understand the investigative nature of science.
4. Have attitudes about and appreciations of scientists, science, and the consequences of science that stem from adequate understandings.

Some of the sub-objectives under the secondary primary objective are:

1. Identify and define a scientific problem.
2. Suggest a scientific hypothesis.
3. Propose validating procedures.
4. Obtain requisite data.
5. Interpret data.
6. Check the logical consistency of hypotheses with relevant observations and experiments (7).

Recently, considerable concern has been manifested about the teaching of the science processes of thought as well as subject matter. The National Science Teachers Association in their publication *Theory Into Action* outlined some of the main themes students should learn about the processes of science as follows:

I. Science proceeds on the assumption, based on centuries of experience, that the universe is not capricious.
II. Scientific knowledge is based on observations of samples of matter that are accessible to public investigation in contrast to purely private inspection.
III. Science proceeds in a piecemeal manner, even though it also aims at achieving a systematic and comprehensive understanding of various sectors or aspects of nature.

IV. Science is not, and will probably never be, a finished enterprise, and there remains very much more to be discovered about how things in the universe behave and how they are interrelated.
V. Measurement is an important feature of most branches of modern science because the formulation as well as the establishment of laws are facilitated through the development of quantitative distinctions (6, p. 20).

The National Science Teachers Association has also listed the major scientific themes students should come to know. These conceptual schemes for the organization of the curriculum (kindergarten to college) are:

I. All matter is composed of units called fundamental particles; under certain conditions these particles can be transformed into energy and vice versa.
II. Matter exists in the form of units which can be classified into hierarchies of organizational levels.
III. The behavior of matter in the universe can be described on a statistical basis.
IV. Units of matter interact. The bases of all ordinary interactions are electromagnetic, gravitational, and nuclear forces.
V. All interacting units of matter tend toward equilibrium states in which the energy content (enthalpy) is a minimum and the energy distribution (entropy) is most random. In the process of attaining equilibrium, energy transformation or matter transformation or matter-energy transformation occur. Nevertheless, the sum of energy and matter in the universe remains constant.
VI. One of the forms of energy is the motion of units of matter. Such motion is responsible for heat and temperature and for the states of matter: solid, liquid, and gases.
VII. All matter exists in time and space and, since interactions occur among its units, matter is subject in some degree to changes with time. Such changes may occur at various rates and in various patterns (6, p. 21).

The National Assessment Program

The National Assessment Program was first started by the Carnegie Foundation to provide information regarding achievement of students in the American school system. The Foundation brought together outstanding educators to outline goals and objectives for several academic areas including science. These are to be used as the basis for designing evaluational instruments to determine student achievement. Outlined below is a preliminary listing of their goals.

I. *Know Fundamental Facts and Principles of Science*

A. Charge, momentum, mass-energy and the conservation of these quantities.

• • • • • •

B. Characteristics of electricity and magnetism.
C. Characteristics of electromagnetic radiation.
D. Atomic and molecular nature of matter.
E. Discontinuities of matter.
F. Classical mechanics (statics, dynamics, mechanics of continuous media).

• • • • • •

G. Heat and simple kinetic theory.
H. Nature (and control) of chemical reactions.
I. Relation of structure of matter to behavior.
J. Common types of chemical behavior.
K. Minerals and rocks.
L. Erosion and weathering.
M. Land forms.
N. Fossils and geological history.
O. Internal constitution of the earth.
P. Diversity of living things.
Q. Organization of living matter.
R. Metabolism.

• • • • • •

S. Behavior of organisms.
T. Reproduction.
U. Evolution.

• • • • • •

V. Ecology.
W. Elements of atmospheric science.
X. Solar system — its arrangement and dynamics.
Y. Stars and galaxies.
Z. Ages and dimensions within the universe — sense of scale.
AA. Health and nutrition.
AB. Nature and properties of matter.
BB. Laboratory observations and techniques in chemistry.

II. *Possess the Abilities and Skills Needed to Engage in the Processes of Science*

A. Ability to identify and define a scientific problem.

B. Ability to suggest or recognize a scientific hypothesis.

• • • • • •

C. Ability to propose or select validating procedures (both logical and empirical).

• • • • • •

D. Ability to obtain requisite data.

• • • • • •

E. Ability to interpret data; i.e., to comprehend the meaning of data and recognize, formulate, and evaluate conclusions and generalizations on the basis of information known or given.

• • • • • •

F. Ability to check the logical consistency of hypotheses with relevant laws, facts, observations, or experiments.

• • • • • •

G. Ability to reason quantitatively and symbolically.

• • • • • •

H. Ability to distinguish among fact, hypothesis, and opinion; the relevant from the irrelevant; and the model from the observations the model was derived to describe.

• • • • • •

I. Ability to read scientific materials critically.

• • • • • •

J. Ability to employ scientific laws and principles in familiar or unfamiliar situations.

• • • • • •

III. *Understand the Investigative Nature of Science*

A. Scientific knowledge develops from observations and experiments and the interpretation of the observations and the experimental results; such observations and experiments are subject to critical examination and to repetition.

• • • • • •

B. Observations are generalized in laws.

• • • • • •

C. Laws are generalized in terms of theories.

• • • • • •

D. Some questions are amendable to scientific inquiry and others are not.

• • • • • •

E. Measurement is an important feature of science because the formulation as well as the establishment of laws are facilitated through the development of quantitative distinctions. Measurements are inherently and only approximate and are progressively inclusive and precise.

• • • • • •

F. Science is not, and will probably never be, a finished enterprise.

• • • • • •

IV. *Have Attitudes About and Appreciations of Scientists, Science, and the Consequences of Science that Stem from Adequate Understandings.*

A. Recognize the distinction between science and its applications.

• • • • • •

B. Have accurate attitudes about scientists.

• • • • • •

C. Understand the relationship between science and misconceptions or superstitions.

• • • • • •

D. Be ready and willing knowingly to apply and utilize basic scientific principles and approaches, where appropriate, in everyday living.

• • • • • •

E. Be independently curious about and participate in scientific activities (1, 9–25).

• • • • • •

Summary

Before behavioral objectives for a course or a curriculum can be defined, goals of instruction must be determined. National organizations and leading educators have invested considerable effort in selecting and establishing these for science and mathematics. The goals outlined for science and mathematics include the learning of concepts and principles in a conceptual scheme framework, plus the various mental operational processes involved in problem solving. Problem solving has been analyzed and refined into a set of operations defined in behavioral terms. These problem-solving behaviors in many instances are similar for both science and mathematics, indicating that science and mathematics teachers are in many instances teaching toward similar goals.

Bibliography

1. Committee of Assessing the Progress of Education, *National Assessment of Educational Progress, Science Objectives,* 20 Huron Towers, 2222 Fuller Road, Ann Arbor, Michigan, 1969.

2. Reprinted from *Goals for School Mathematics,* copyright © 1963 by Educational Development Center, Inc. (formerly Educational Services Incorporated). Reprinted by permission of Houghton Mifflin Company.

3. *Goals for the Correlation of Elementary Science and Mathematics.* The Report of the Cambridge Conference on the Correlation of Science and Mathematics. Boston, Massachusetts: Houghton Mifflin Company, 1969.

4. Husen, T., ed. *International Study of Achievement in Mathematics — A Comparison of Twelve Countries Volume 1.* New York: John Wiley & Sons, 1967.

5. Johnson, D. and Rising, G.R. *Guidelines for Teaching Mathematics.* Belmont, California: Wadsworth Publishing Company, Inc., 1967, p. 11.

6. National Science Teachers Association Curriculum Committee. *Theory Into Action in Science Curriculum Development.* National Science Teachers Association, Washington, D.C., 1964.

7. Thompson, R.E. "Development of Science Objectives and Exercises for National Assessment." Unpublished paper, Sixteenth Annual Convention, National Science Teachers Association, Washington, D.C., April 1, 1968.

CHAPTER THREE

PROBLEM-SOLVING BEHAVIORAL OBJECTIVES

Instructional Model

In writing behavioral objectives, an instructor needs to determine his goals, as outlined in the previous chapters, and the subject matter content related to these goals. In any instruction, there are generally three components, as indicated by the chart below.

Subject matter	Critical thinking	Feelings, values and attitudes about subject matter

When prospective and experienced teachers start to write behavioral objectives, they emphasize mainly only the first component — learning subject matter. It should be clear, however, that subject matter objectives can be written so that the second component — critical thinking — is integrated with them. Compare the two objectives below which include subject matter emphases only.

The student should be able to:

1. *List* the limiting adaptive factors in the desert environment.
2. *Hypothesize* what factors of the desert environment would limit the population of a turtle introduced into it.

Note that the first objective mainly demands only knowledge. The second objective, however, requires the student to reason and use a cognitive mental process of the scientific method, hypothesize, plus apply his knowledge. If the above objectives were written by different teachers, it is obvious that the second instructor is endeavoring to teach for higher levels of learning compared to the first teacher.

In writing behavioral objectives, an instructor must constantly guard against stressing only one goal of instruction — knowledge of subject matter to the detriment of the other goals. One way to do this is to keep a list of problem-solving behaviors constantly before you and review your objectives after writing them to see if you are integrating these with subject matter to the best of your ability.

Problem-Solving Objectives

Problem-solving behaviors have been defined through the efforts of outstanding science and mathematics educators over several years. The U.S. Office of Education has published a list of problem-solving behaviors prepared by Dr. Darrell Barnard and Dr. Ellsworth Obourn. This is one of the best sources for specific problem-solving behavioral objectives. The list appears below, slightly modified by us in behavioral terms.

I. *Attitudes which can be developed through science teaching.* The science program should develop the attitude which will modify the individual's behavior so that:

A. When confronted with natural phenomena, he:

1. States he does not believe in superstitions such as charms or signs of good or bad luck, nor does he carry them.
2. States that occurrences which seem strange and mysterious can usually be explained finally by natural cause.
3. Describes why a connection between two events, because they occur at the same time, cannot necessarily be made.
4. States he believes that truth never changes, but that his ideas of what is true may change as he gains better understanding of the truth.
5. Describes how he bases his ideas upon the best evidence and not upon tradition alone.
6. Revises his opinions and conclusions in light of additional reliable information.
7. Listens to, observes, or reads evidence supporting ideas contrary to his personal opinions.

8. Accepts no conclusion as final or ultimate.

B. When confronted with a problem, he bases opinions and conclusions on adequate evidence, and he:

1. Is slow to accept as facts any that are not supported by convincing proof.
2. Bases his conclusions upon evidence obtained from a variety of dependable sources.
3. Hunts for the most satisfactory explanation of observed phenomena that the evidence permits.
4. Sticks to the facts and refrains from exaggeration.
5. Does not permit his personal pride, bias, prejudice, or ambition to prevert the truth.
6. Does not make snap judgments or jump to conclusions.

C. In problem solving, he evaluates techniques and procedures used and information obtained, and he:

1. Uses a planned procedure in solving his problems.
2. Uses the various techniques and procedures which may be applied in obtaining information.
5. Adapts the various techniques and procedures to the problem at hand.
4. Personally considers the information obtained and states whether it relates to the problem.
5. Judges and states whether the information is sound, sensible, and complete enough to allow a conclusion to be made.
6. Selects the most recent, authoritative, and accurate information related to the problem.

D. When confronted with a problem, he is curious concerning the things he observes, and:

1. Asks "whys," "whats," and "hows" of observed phenomena.
2. States he is not satisfied with vague answers to his questions.

II. *Problem-solving abilities which can be developed through science teaching.*

The science program should develop those abilities involved in problem solving which will modify the individual's behavior. The student should be able to:

A. Formulate significant problems:

1. Describe situations involving personal and social problems.
2. Describe specific problems in these situations.
3. Isolate the single major idea in the problem.
4. State the problem in question form.
5. State the problem in definite and concise language.

B. Analyze problems:

1. Pick out the key words of a problem statement.
2. Define key words as a means of getting a better understanding of the problem.

C. Obtain information regarding a problem from a variety of sources:

1. Recall past experiences which bear upon his problem.
2. Isolate elements common in experience and problem.
3. Locate source materials.
   a. Use the various parts of a book.
      1. Use the key words in the problem statement for locating material in the index.
      2. Choose proper sub-topics in the index.
      3. Use alphabetical materials, cross references, the table of contents, the title page, the glossary, figures, pictures and diagrams, footnotes, topical headings, running headings, marginal headings, an appendix, a pronunciation list, and "see also" references.
   b. Use materials other than textbooks, such as, encyclopedias, popularly written books, handbooks, dictionaries, magazines, newspapers, pamphlets, catalogues, bulletins, films, apparatus, guide letters, numbers, signs marking the location of information, bibliographies, etc.
   c. Use library facilities, such as, the card index, the Readers' Guide, and the services of the librarian.
4. Use source materials.
   a. Use aids in comprehending material read.
      1. Find main ideas in a paragraph.
      2. Use reading signals.
      3. Formulate statements from reading.
      4. Phrase topics from sentences.
      5. Skim for main ideas.
      6. Learn meanings of words and phrases from context.

7. Select the printed material related to the problem.
8. Cross-check a book concerning the same topic.
9. Recognize both objective and opinionated evidence.
10. Determine the main topic over several paragraphs.
11. Take notes.
12. Arrange ideas in an organized manner.
13. Make outlines.

b. Interpret graphic material.
    1. Obtain information from different kinds of graphic material.
    2. Read titles, column headings, legends, and data recorded.
    3. Formulate the main ideas presented.
    4. Evaluate conclusions based upon the data recorded.

5. Use experimental procedures appropriate to the problem.
    a. Devise experiments suitable to the solution of the problem.
        1. Select the main factor in the experiment.
        2. Allow only one variable.
        3. Set up a control for the experimental factor.
    b. Carry out the details of the experiment.
        1. Identify effects and determine causes.
        2. Test the effects of the experimental factor under varying conditions.
        3. Perform the experiment for a sufficient length of time.
        4. Accurately determine and record quantitative and qualitative data.
        5. Develop a logical organization of recorded data.
        6. Generalize upon the basis of organized data.
    c. Manipulate the laboratory equipment needed in solving the problem.
        1. Select kinds of equipment of materials that will aid in solving the problem.
        2. Manipulate equipment or materials that will aid in an understanding of its function to the outcome of the experiment.
        3. Recognize that equipment is only a means to the end results.
        4. Determine the relationship between observed actions or occurrences and the problem.
        5. Appraise scales and divisions of scales on measuring devices.
        6. Obtain correct values from measuring devices.
        7. State the capacities of limitations of equipment.

8. Return equipment clean and in good condition.
9. Avoid hazards and consequent personal accidents.
10. Practice neatness and orderliness.
11. Avoid waste in the use of materials.
12. Exercise reasonable care of fragile or perishable equipment.

6. Solve mathematical problems necessary in obtaining pertinent data.
   a. Pick out the elements in a mathematical problem that can be used in its solution.
   b. State relationships between these elements.
   c. Use essential formulae.
   d. Perform such fundamental operations as addition, subtraction, multiplication, and division.
   e. Use the metric and English systems of measurement.
   f. Understand the mathematical terms used in these problems; i.e., square, proportion, area, volume, etc.
7. Make observations suitable for solving the problem.
   a. Observe demonstrations.
      1. Devise suitable demonstrations.
      2. Select materials and equipment needed in the demonstration.
      3. Identify the important ideas demonstrated.
   b. Pick out the important ideas presented by pictures, slides, and motion pictures.
   c. Pick out the important ideas presented by models and exhibits.
   d. Use the resources of the community for purposes of obtaining information pertinent to the problem.
      1. Locate conditions of situations in the community to observe.
      2. Pick out the essential ideas from such observation.
8. Use talks and interviews as sources of information.
   a. Select individuals who can contribute to the solution of the problem.
   b. Make suitable plans for the talk or interview.
   c. Appropriately contact the person who is to talk.
   d. Select the main ideas from the activity.
   e. Properly acknowledge the courtesy of the individual interviewed.

D. Organize the data obtained:

1. Use appropriate means for organizing data.
   a. Construct tables.

b. Construct graphs.
c. Prepare summaries.
d. Make outlines.
e. Construct diagrams.
f. Use photographs.
g. Use suitable statistical procedures.

E. Interpret organized data:

1. Select the important ideas related to the problem.
2. Identify the different relationships which may exist between the important ideas.
3. State these relationships as generalizations which may serve as hypotheses.

F. Test the hypotheses:

1. Check proposed conclusion with authority.
2. Devise experimental procedures suitable for testing the hypotheses.
3. Recheck data for errors in interpretation.
4. Apply hypothesis to the problem to determine its adequacy.

G. Formulate a conclusion:

1. Accept the most tenable of the tested hypotheses.
2. Use this hypothesis as a basis for generalizing in terms of similar problem situations (1).

Science Process Objectives

Dr. Robert S. Tannenbaum has outlined in behavioral terms the science processes students should develop in grades 7–9. This list was compiled from a study of relevant texts and other literature as a basis for constructing a standardized test entitled *Test of Science Processes* for the aforementioned grades. The processes were validated as important and desirable for the junior high school by 31 experts from throughout the United States. The list appears below.

Evaluating Science Processes

Process I: *Observing*

In order for a student to demonstrate competence in using the process of observing, he should be able to do the following:

1. Demonstrate an operational knowledge of the properties of objects and systems, e.g., phase, motion, color, mass, length, volume, area, and temperature, and the rate of change of any of the above.
2. Describe objects primarily in terms of their properties rather than in terms of their uses avoiding anthropomorphic descriptions.
3. Identify and describe the objects which interact in a system.
4. Identify and describe the interactions of objects and systems of objects.
5. Analyze an object in terms of an appropriate system of component parts and list and describe those parts.
6. Distinguish among various spatial relationships of the objects within a particular system and describe systems in terms of the spatial relationships of their members.
7. List in detail the observable characteristics of a given phenomenon.

Process II: *Comparing*

In order for a student to demonstrate competence in using the process of comparing, he should be able to do the following:

1. Describe the similarities and differences in terms of physical properties of two or more of each of the following:
   a. Objects.
   b. Systems of objects.
   c. Interactions of objects and of systems of objects.
   d. Relative positions of objects.
2. Contrast two or more of each of the following on the basis of their properties:
   a. Objects.
   b. Systems of objects.
   c. Interactions of objects and of systems of objects.
   d. Relative positions of objects.

Process III: *Classifying*

In order for a student to demonstrate competence in using the process of classifying, he should be able to do the following:

1. Group objects or systems of objects according to a given property.
2. Select and justify an appropriate property and group objects or systems of objects according to that property.
3. Group objects or systems of objects according to two or more given simultaneous properties. (Simultaneous properties are ones which are to be considered together, e.g., objects which are at the

same time red and square vs. objects which are not at the same time red and square.)

4. Select and justify two or more appropriate simultaneous properties and group objects or systems of objects according to those properties.
5. Given a group of objects, identify those properties on which they are grouped.
6. Given a set of objects or systems of objects, remove a specified number of members from the original set to form two net sets which are grouped on the basis of one or more given simultaneous properties.

Process IV: *Quantifying*

In order for a student to demonstrate competence in using the process of quantifying, he should be able to do the following:

1. Demonstrate an operational knowledge of a set, the elements of a set, the union of two or more sets, and the intersection of two or more sets.
2. Demonstrate an operational knowledge of ordinal and cardinal numbers up to one million and of negative numbers.
3. Demonstrate an operational knowledge of simple fractions, percents, and decimals.
4. Be able to order a group of objects or systems of objects from most to least (or vice-versa) on one of more simultaneous orderable properties.
5. Be able to arrange and to read data in various graphic and tabular formats.

Process V: *Measuring*

In order for a student to demonstrate competence in using the process of measuring, he should be able to do the following:

1. Suggest and use "home-made" units for measuring the properties of objects.
2. Demonstrate an operational knowledge of units of measure, the function of widely accepted units, the names and approximate sizes of the most common units such as inch, foot, centimeter, meter, pound, quart, gram, kilogram, liter, second, degree, Celsius, etc.
3. Be able to select the appropriate units for making a particular measurement.
4. Be able to *estimate* the dimensions and properties of an object (including temperature, size, mass, etc.) for purposes of ordering, describing, and classifying.

5. Be able to *measure* the dimensions and properties of an object (including temperature, size, mass, etc.) for purposes of ordering, describing, and classifying.
6. Demonstrate an operational knowledge of area and volume in terms of one, two, and three dimensional measurements (e.g., $a = 1^2$ and $a = 1 \times w$; $v = 13$, $v = h \times a = h \times 1^2$, and $v = h \times 1 \times w$).
7. Be able to measure time.
8. Be able to measure the rate of change of a property of an object or a system of objects.
9. Represent and recognize an object or a system of objects by a scale diagram.
10. Represent and recognize the spatial relationships among two or more objects or systems of objects by a scale diagram (mapping).
11. Recognize the appropriateness and limitations of measuring devices in a given situation.

Process VI: *Experimenting*

In order for a student to demonstrate competence in using the process of experimenting, he should be able to do the following:

Use suitable experimental procedures in seeking solutions to problems including possibly:

1. Design an investigation appropriate to the problem:
   a. Select, clarify, and state in testable terms (perhaps as an answerable question) the primary variable to be investigated.
   b. Limit the number of variables to a workable number.
   c. Control the variables appropriately so that logical conclusions may be drawn with regard to the primary variable.
   d. Distinguish between dependent and independent variables.
2. Perform the investigation:
   a. Design, construct, or select, and successfully utilize apparatus to assist in data gathering, where appropriate.
   b. Employ the processes of observing, comparing, quantifying, classifying, and measuring to gather data.
   c. Repeat the data gathering a sufficient number of times to improve reliability and under sufficiently varying conditions to account for the influence of different variables.
   d. Record and organize the data gathered in a logical form.
3. Utilize the processes of inferring and predicting to interpret the data collected, answer the original problem, and lead to the posing of new problems and the design of new experiments to investigate them.

Process VII: *Inferring*

In order for a student to demonstrate competence in using the process of inferring, he should be able to do the following:

1. Draw warranted generalizations from a body of data.
2. Identify the factor most likely to have caused a given change in a system.
3. Identify and specify the observations which would be needed to justify a particular generalization.
4. Be able to distinguish between a statement based directly on observations and one which is an inference or a generalization.
5. Be able to draw more than one inference in situations where the data allows this.
6. Be able to test an inference by collecting further data.
7. Recognize which data are necessary and sufficient to support an inference of a generalization.

Process VIII: *Predicting*

In order for a student to demonstrate competence in using the process of predicting, he should be able to do the following:

1. Be able to detect or demonstrate trends in data (presented in many different ways) and be able to use these trends to predict by extrapolation and/or interpolation.
2. Devise and use simple means of checking the accuracy of the predictions made.
3. Recognize and use pertinent arguments, reasons, or principles to justify a prediction.
4. Demonstrate an operational knowledge of the necessity for multiple and reliable observations prior to prediction and an unwillingness to offer predictions in the absence of such observations (2).

Summary

Although problem solving has been a major instructional aim for decades, it seldom received much attention in the traditional classroom. This probably has been largely due to the mistaken belief by teachers that if they taught science or mathematics, the problem-solving potential of their students would naturally unfold. Educators have become increasingly concerned over the last decade about the ineffective methods of teaching problem solving in our schools. There have been, as a result, tremendous national efforts to revise the curriculum and train teachers to more easily reach this goal. Today, it is realized that students become proficient as problem solvers only if they have numerous opportunities to solve problems. It is for this reason that the new science and mathematics projects stress the investigative approach to learning.

Teachers probably haven't taught for problem solving because they haven't known how to teach for it, nor what kind of behaviors to elicit. The work done by Barnard, Obourn, Tannenbaum, and others in describing these problem-solving behaviors should help considerably in overcoming this obstacle. If students are involved in formulating and analyzing problems, hypothesizing, designing investigations, collecting, interpreting data, and making conclusions as suggested above, then they are behaving in a problem-solving manner. Stating problem-solving behavioral objectives, designing activities for their manifestation, and evaluating student achievement relative to their attainment gives students better insights into the nature of science and mathematics, while increasing their intellectual abilities. How well a teacher carries out these functions determines to a large extent the degree of insight that the students will attain.

Bibliography

1. Obourn, Ellsworth and Bernard, J. Darrell. *Science Teaching Service Circular*. An analysis and Checklist on the Problem Solving Objective. U.S. Office of Education, Washington, D.C., forthcoming.
2. Tannenbaum, Robert S. "The Development of the Test of Science Processes." *Journal of Research in Science Teaching*, forthcoming.

CHAPTER FOUR

THE WRITING OF BEHAVIORAL OBJECTIVES

The following steps are necessary to consider in starting to write objectives:

1. Select goals.
2. Select content.
3. Write a tentative statement describing how you expect the student to perform with the given content.
4. Rewrite your statements in terms of performance criteria. Be sure to include:
   a. A description of the behavior expected of the student.
   b. The conditions under which the student must exhibit the given behavior.
   c. The criteria of success.
5. Analyze and evaluate your objectives. Be sure you have included higher cognitive levels.

Select Goals

Chapter two outlines the goals of science and mathematics. Choose from these lists the ones you think are most important. Add any others you think are necessary.

Select Content

As a first step in writing behavioral objectives, instructors should compile lists of content they wish to communicate. Textbooks, curriculum guides, and professional journals are excellent sources for this activity. Pre-arranged lists may be helpful, but the final decision of the most "worthwhile" content should be made by a curriculum specialist and teachers at the school level. "Unless educators at the local and state level are willing to make the effort . . . someone else, probably commercial producers of package instructional programs, will" (2, p. 57).

Tentative Statements of Behavioral Objectives

The following form is suggested to use initially in starting to write behavioral objectives:

The student should be able to:

1. Hypothesize what will happen to a slug when placed in salt water (hypertonic solution).
2. Design an experiment to determine whether coffee or tea retains heat better.
3. Infer that slight environmental changes may cause the death of certain organisms, for example, the slug.

The objectives below are written in a poorer style:

1. The student should circle the numeral representing the number of triangles in a given diagram.
2. The student should join two sets by drawing a single loop around them.
3. The student should indicate the number of objects in each of two sets.

Compare these two sets of objectives. Why is the second set in a poorer format than the first? Note that the phrase "the student should be able to" is written once and only once in the first set. The second set uses the statement over and over again. This forces the reader to read the same expression several times. The reader is mature enough to hold this phrase in his mind after the initial reading. Using the statement once economizes the reader's time.

Note also that the action verbs, *hypothesize, design,* and *infer,* are underlined in the first set. It is our experience that placing the action verb first and underlining it defines the behaviors more accurately. Teachers, as a result, are usually able to determine more easily if a statement is in behavioral terms. Compare the following objectives:

1. Defines acceleration operationally.
2. Understands the law of conservation of energy.
3. Knows how to use it.
4. Identifies symbols used in the first year calculus, e.g. $\frac{dy}{dx}$, $f'(x)$.

Which of these are written behaviorally? How does the underlining help you identify the behavior?

Performance Description

The verb of an objective should be the behavior or a description of the performance expected of the successful learner. These are called action verbs. For example, *identifies, names* and *infers* are action verbs, while knows or understands is not.

It has been thought by some educators that the categories used in Bloom's *Taxonomy* could serve as a source of action verbs. This is not the case. However, the categories can be modified for this purpose as shown below.

Categories of Intellectual Behavior

A. *Knowledge* and information of definitions, notation, operations and mathematical concepts.
B. *Techniques and skills* in manipulation and computation.
C. *Translation* of data into symbols and *interpretation* of symbolic data.
D. *Comprehension* to follow reasoning, to read and understand new mathematical developments, and to analyze problems and determine the operations which may be applied to them.
E. *Inventiveness* to think creatively in mathematics, to construct proofs and to apply mathematical ideas and relationships to new problems (5. p. 11).

A detailed set of performance descriptions has been adopted by the American Association for the Advancement of Science, *Science — A Process Approach* Curriculum Project. Each behavior or "action verb" is defined as follows:

Identifying — The individual selects (by pointing to, touching, or picking up) the correct object of a class, in response to a class name. For example, upon being asked, "Which animal is the frog?" when presented with a set of small animals, the child is expected to respond by picking up, or clearly pointing to, or touching the frog.

Distinguishing — Identifying under conditions in which the objects or events are potentially confusable (a square and a rectangle whose lengths are almost equal), or when two contrasting identifications (such as, right and left) are involved.

Constructing — Generating a construction or drawing which identifies a designated object or set of conditions. For example, beginning with a line segment, the request is made to "complete this figure so that it represents a triangle."

Naming — Supplying the correct name (orally or in writing) for a class of objects or events. For example, when asked, "What is this three-dimensional object called?" to respond, "A cone."

Ordering — Arranging two or more objects or events in proper order in accordance with a stated category. For example, "Arrange these moving objects in order of their speeds."

Describing — Generating and naming all of the necessary categories of objects, object properties, or event properties that are relevant to the description of the designated situation. For example, upon being asked to describe this object, the child should respond in sufficient detail so that there is a probability of approximately one that any other individual who hears the description will be able to identify the object or event.

Stating a Rule — The child makes a verbal statement which conveys a rule or principle (the language used to convey the rule or principle need not always be in technical terms, but must include the names of the proper classes of objects or events in their correct order). For example, upon being asked, "What is the test for determining whether this surface is flat?" the child would respond by mentioning the application of a straightedge, in various directions, and to determine whether the straightedge is touching all along the edge.

Applying a Rule — Using an acquired rule or principle to derive an answer to a question. The answer may be a correct identification, the

> supplying of a name, or the construction of a figure. For example, when asked, "Which of the following figures satisfies the definition of an angle?" the child's response should identify two rays with a common end point.
>
> *Demonstrating* — Performing the operations necessary to the application of a rule or principle. For example, if the child is asked to "show me how you would tell whether this surface is flat," then the response requires that the individual use a straightedge, determining whether the edge touches the surface at all points along the edge, and repeat this touching in various directions.
>
> *Interpreting* — The child should be able to identify objects and/or events in terms of their consequences. There will be a set of rules or principles always connected with this behavior (1, pp. 3–4).

The above descriptions have been arranged in a hierarchy from recall (identifying, naming) to application (applying a rule, demonstrating).

While any set of action verbs may be used, the following are most useful for science and mathematics education:

1. Knowledge
   a. Identifies
   b. Names
   c. Chooses
   d. Lists
   e. Selects
   f. Distinguishes
2. Comprehension
   a. Computes
   b. Measures
   c. Matches
   d. Demonstrates
   e. Selects instruments
   f. Balances
3. Application
   a. Compares
   b. Groups
   c. Arranges
   d. Calibrates
   e. Dissects
   f. Operates
4. Analysis
   a. Selects hypotheses
   b. Estimates
   c. Interrelates
   d. Limits
   e. Infers
   f. Reflects
5. Synthesis
   a. Proves
   b. Extrapolates
   c. Interpolates
   d. Predicts
   e. Infers
   f. Deduces

6. Evaluation
   a. Controls variables
   b. Rejects
   c. Verifies
   d. Questions
   e. Interprets
   f. Doubts

Although teachers generally think of performance in terms of pencil and paper, behavior may be defined by specifying the way a learner is to act in other ways. A behavior may be defined as the process of demonstrating the ability to use a scientific tool or a mathematical formula. For example, using a magnet to attract or repel objects. A child who is able to perform this activity should be credited with understanding magnetism independently of his ability to verbalize that understanding.

Write the Conditions

Detailed objectives should stipulate the conditions under which the student must exhibit the desired behavior. The conditions indicate the clues or stimuli presented to the student. They also place restrictions on the student at the time the terminal behavior is sampled. The teacher may identify necessary conditions for stating objectives by asking:

1. What aids will the student be permitted to use?
2. What aids will the student be denied?
3. In what context or situation will the behavior occur?
4. What skills or behaviors should be specifically avoided? Are these excluded by the objective?

Examples of conditions used in objectives are as follows:

1. When given a protractor
2. Given a seed and some water
3. Without the use of a multiplication table
4. Given a chemistry-physics handbook
5. When presented with distillation apparatus

Write the Extent of the Achievement

A criteria of success must be given to determine if the student has achieved the objective. The teacher can indicate the extent of performance by including the response to the following questions in the objective:

1. How often must the learner exhibit the terminal behavior?

2. Is the student with ten correct responses more successful than the one with five correct responses?
3. Has the student who lists three out of five possible responses achieved the objective?
4. Is student X more successful because he responds more quickly than student Y?

Some writers of objectives establish minimum acceptable performance in one or more of the following ways:

1. *Time limit.* The writer specifies that the student must respond or complete the task within a given amount of time. An example would be the statement, "The student must be able to correctly solve at least seven linear equations within a period of thirty minutes (7, p. 45)." Another would be, "The student must weigh a given sample on a balance to within 1/100th of a milligram within one minute." This practice is questionable from an educational viewpoint. Time limits should vary with ability. Slow learners should be given ample time to meet the minimum acceptable performance levels.
2. *Number of Correct Responses.* Acceptable performance may also be defined by specifying the number of correct responses. An example of this method is programmed material where the learner paces himself or is paced by a computer. The learner is permitted to proceed from one unit to the next after he answers correctly all the items in the end-of-unit test or after he correctly responds to 90 per cent of the frames in the unit. The percentage of correct responses necessary for success is arbitrary and varies with the difficulty of content and the size of the steps in the program..
3. *Standardized Test Comparison.* Another means of establishing minimum performance levels is to compare the child with a nation-wide sample of children. This occurs when a child is ranked according to norms on standardized tests. Minimum performance is determined by the norm for children his age or at his grade level. This practice is also being disputed as theories of learning raise questions concerning readiness and rate of development of knowledge.

   The extent may not be specifically stated in some objectives; then it should be assumed the student will exhibit the behavior whenever he encounters the appropriate situation.

A Sample Objective

The complete statement of an objective should include the performance description, the conditions under which the learner must exhibit the behavior, and the criteria for success. The statement, "The child should be able to *name* the area of a square," contains the behavior *naming* and the content *area of a square*. By supplying appropriate conditions and extent, the statement may be modified into a meaningful behavioral objective for various grade levels. At the specified grade level, the student should be able to:

	First Grade
Performance Description	*Name* the area of a large square by filling it in with unit squares and counting the number of unit squares.
Condition	He will be asked how many of the smaller squares fill up the larger one without overlapping.
Extent	He must give the correct response whenever presented with a square whose side is not more than six units long.

	Third Grade
Performance Description	*Name* the area of a square by measuring the length of a side of the square and computing the area.
Condition	He will be presented with a drawing of an unmarked square and a ruler and asked, "What is the area of the square in square inches?"
Extent	The measurement must be accurate to within one-quarter of an inch.

	Sixth Grade
Performance Description	*Name* the area of a square by writing the correct symbolic representation for the area of the square in two different ways.
Condition	He will be given a square, marked on each side by a letter and asked for the area of the square.
Extent	He must respond correctly at least eight times when presented with ten squares with a variety of letters (A, X, Z, etc.) marking their sides.

How specific should the statement of an objective be? Maurice L. Hartung, in the 26th Yearbook of the National Council of Teachers of Mathematics states:

> Objectives should be formulated to a level of specificity such that it is possible readily to infer some learning activities appropriate for helping students achieve each objective, and also to devise means of evaluating the achievement, but not to a greater level of specificity than is needed for these purposes (4, p. 30).

Note that the sample objectives above satisfy both of these conditions.

Technique for Analyzing and Evaluating Objectives

A behavior-content grid is a device for visualizing relationships between content and behavior. It may be used in the selection, formulation, and evaluation of behavioral objectives. Its use in constructing tests and measuring and reporting achievement will be discussed in chapter eight. The example below has been constructed for a course in geometry. It may be refined by including sub-categories of content and behavior.

BEHAVIORS

CONTENT	Knowledge of Definitions and Terms	Skills and Techniques in Manipulation & Computation	Translation and Interpretation	Analysis-Select Relevant and Irrelevant Data	Follow and Construct Proofs	Verify and Criticize
Non-Metric Plane Geometry	√					
Measurement of Length and Angle	√	√				
Simple Closed Curves	√					
Circles	√		√	√		
Constructions	√	√				
Congruent Sets of Points	√					
Area of Plane Regions	√	√	√			
Similar Figures						
Graphing Points in a Plane	√	√				

To construct a behavior-content grid, do the following: make a list of the topics to be covered; list the behaviors you think are important; arrange the content along one side of a rectangular grid and the behaviors along the other. The intersection of a row and a column is called a cell. A check in one of these cells represents an objective involving a particular behavior with a specific piece of content. Every piece of content will not necessarily interact with every behavior. Factual content, for example, usually requires only low level cognitive behaviors, such as, recall or comprehension.

A set of objectives may be analyzed by locating each objective in one of the cells of the grid. Now, look at the distribution of the checks on the grid. Are they evenly distributed or clustered? Are there many in one column and few in the others? Do any behaviors have no objectives associated with it? Does any content area have no objectives?

In the example on page 45, notice the heavy emphasis on knowledge of definitions and terms and the lack of objectives requiring higher cognitive behaviors. This is a common situation for initial attempts at writing objectives because low level ones are easier to write. This situation is also common in programs using older curriculum materials.

Notice also that the writer has included several objectives on circles and none on similar figures. Analysis of a test in this way reveals a preference for certain topics. It establishes the content considered important and unimportant by the writer.

If you haven't already done so, try writing a set of objectives. A previously written set will do. Now analyze the set with a behavior-content grid. Have you neglected any topics or behaviors? Have you placed too much emphasis on memory and recall? Are additional objectives required? What does this analysis reveal to you about your teaching?

Materials on Writing Behavioral Objectives

Among the many good sources on writing behavioral objectives are:

1. *Preparing Instructional Objectives,* Robert F. Mager (Palo Alto, Calif.: Fearon Publishers 1962).

   Mager's book is compact and entertaining. It is programed and includes self-evaluation pages to help the reader evaluate his progress. Although the book contains few examples appropriate for science and mathematics, it is easy to read and a good beginning point for learning how to construct behavioral objectives.

2. *Behavioral Objectives and Instruction*, Robert J. Kilber, Larry L. Baker, and David T. Miles (Boston, Mass.: Allyn and Bacon, Inc. 1970).

   This book is a general text listing few examples from science and mathematics, but is broad in scope, giving several excellent suggestions.

3. *Stating Behavioral Objectives for Classroom Instruction*, Norman E. Gronlund (London: The Macmillan Co. 1970).

   This small book has many practical features including a check list for evaluating objectives and an appendix of action verbs.

4. A series of filmstrips with audio tapes by Vimcet Associates, P. O. Box 24713, Los Angeles, Calif. 90024.

   The filmstrips are titled:

   a. *Selecting Appropriate Educational Objectives*
   b. *Educational Objectives*
   c. *Establishing Performance Standards*
   d. *Perceived Purpose*
   e. *Appropriate Practice*
   f. *Systematic Instructional Decision Making*
   g. *Evaluation*

   In addition to the audio tape, there is a written guide accompanying each presentation. At appropriate places, the student is asked to demonstrate his comprehension by answering questions relating to the presentation. In this way, the student is given immediate feedback.

A Source of Behavioral Objectives

Writing good behavioral objectives is a time consuming job. Most classroom teachers don't have the time for writing objectives and carrying out instruction. In response to this need, Dr. W. Jones Popham has established an "Instructional Objectives Exchange" (IOX) at the U.C.L.A. Center for the Study of Evaluation. The purpose of the Objectives Exchange is to make available alternate objectives and measuring devices. These objectives and measures will assist school personnel in instructional and evaluation activities. The Exchange will function in three ways:

1. It will serve as a clearing house through which the nation's schools can exchange instructional objectives, thereby capital-

izing on the developmental efforts of other educators rather than being obliged to commence afresh the development of objectives.

2. It will collect and, when necessary, develop measuring techniques suitable for assessing the attainment of the objectives available through the Exchange.
3. It will develop properly formulated instructional objectives in important areas where none currently exist, that is, fill the gaps not covered by available objectives (8, p. 1).

The Exchange currently has available collections of objectives in mathematics (K-9), biology (10-12), and other non-science areas.

> Each collection contains a) general introductory information, b) an overview of the content of the particular booklet, including c) appropriate credit to the schools, individuals, or other agencies contributing these particular objectives, d) the objectives themselves, one per page, with e) a sample test item per objective. For every objective an estimate of the particular grade level at which the objective might be introduced is provided. A rough indication of the objective's appropriateness for the Exchange is also given (more specifically, a rating regarding whether IOX should make the objective available to Exchange users). Finally, g) a brief questionnaire is provided so that users of the collection can provide feedback data regarding the quality of these objectives (8, pp. 7, 8).

Sample Objective from the Exchange:

MATHEMATICS

MAJOR CATEGORY: Applications, Problem Solving
SUB-CATEGORY: Statistics

GRADE INTRODUCED: 7
IOX ACCEPTABILITY RATING: 2

OBJECTIVE: Given a situation in which a faulty conclusion is presented, the student will analyze the data and identify the errors in logic.

SAMPLE ITEM: What is wrong with the conclusion based on the given data? More people were killed in airplane accidents in 1969 than in 1929.

ANSWER: There were less planes in 1929. It is better to talk about the percentage of deaths in relation to the number of air miles flown in these two years (8, p. 10).

Additional information and a catalog of objective collections may be obtained from:

> Instructional Objectives Exchange
> Box 24095
> University of California
> Los Angeles, California 90024

Summary

The first step in writing behavioral objectives is to select the goals you think are important. Next, choose the content. Write several tentative statements indicating what you expect the student to be like after your instruction. These statements should reflect he goals and content you chose. The verb should describe an observable behavior. Rewrite these statements, being sure to include:

1. Performance description
2. Conditions
3. Extent

Make the wording as precise and unambiguous as possible. Check to see if each statement suggests a learning activity and an evaluation situation. Now, analyze the set of objectives with a behavior-content grid. Finally, modify the set by deleting or adding objectives.

Each set of objectives should include both cognitive and affective objectives. Discussion of objectives related to the affective domain has been delayed to avoid confusion. The subject of affective objectives will be taken up in the next chapter.

Bibliography

1. American Association for the Advancement of Science. Commission on Science Education Newsletter, vol. 1, no. 2, April 1965, pp. 3-4.

2. Gentry, A. *Planning a Continuous Science Program for All Junior High School Youth, Guidelines for Designing and Implementing Contemporary Science Curricula for Grades 7-9*. Riverside, California: Office of the Superintendent of Schools, 1967.

3. Gronlund, N. *Stating Behavioral Objectives for Classroom Instruction.* London: The Macmillan Company, Collier-Macmillan Limited, 1970.

4. Hartung, M. "Basic Principles of Evaluation." *Evaluation in Mathematics, Twenty-sixth Yearbook.* Washington, D.C.: National Council of Teachers of Mathematics, 1961.

5. Keeves, J. "Evaluation of Achievement in Mathematics: The Tests Used and Attainment in Mathematics of Australian Pupils in The I.E.A. Project, 1964." Australian Council for Educational Research, no. 4, series no. 6, 1966.

6. Kilber, R., Baker, L. and Miles, D. *Behavioral Objectives and Instruction.* Boston, Mass.: Allyn and Bacon, Inc., 1970.

7. Mager, R. *Preparing Instructional Objectives.* Palo Alto, California: Fearon Publishers, 1962.

8. Popham, W. *Instructional Objectives Exchange Catalog.* Los Angeles, California: U.C.L.A. Graduate School of Education, 1970.

Chapter Five

THE WRITING OF AFFECTIVE OBJECTIVES

The Hierarchy of Affective Objectives

In addition to a taxonomy of cognitive objectives, there is also one for the affective domain written by Krathwohl, Bloom, and Masia (2). The word affective comes from the Latin word *affectus*, meaning capable of feeling or emotion. The affective objectives are those in which the instructor endeavors to have his students learn something and develop a feeling for what he is learning. These include:

1. Appreciations
2. Feelings
3. Values
4. Attitudes

Whenever an instructor teaches anything, he teaches a feeling about it. For example, if he comes into class and says, "We have to study mathematics now. I know you don't like it, and to tell you the truth, I don't either," he is teaching that mathematics is no fun and is boring. As a result, his students develop negative feelings toward the subject. *The most important thing an educator can do is develop positive emotions toward what is being learned.* If a student learns to enjoy learning he will continue to grow intellectually long after his formal education is through.

Krathwohl, Bloom, and Masia have classified the affective domain into five main groups as follows:

1. *Receiving* (attending). The student is at least willing to hear or study the information.
2. *Responding.* The student will respond about the material being studied.
3. *Valuing.* The student has a commitment to what is being learned and believes it has worth.
4. *Organization.* The student has a hierarchy of value— a value system.
5. *Characterization by a value or value complex.* The student has an internalized value complex directing his total behavior. He has integrated his beliefs, ideas, and attitudes into a philosophy of life (2, Appendix A).

This classification system can be shortened to include three general categories as seen below:

1. Awareness
2. Acceptance of values
3. Preference for values

It is more functional to keep in mind these three general levels and this hierarchy in beginning to write affective objectives than it is to use the more detailed one. All you have to do is ask yourself, "Do I want to determine whether or not my students are aware of a topic, if they value it, or if their total value complex has changed because of the instruction they have had?"

Overt and Covert Affective Achievement

Some affective behavior (overt) can be observed, while some is internalized (covert) and cannot be seen. The latter type can be assessed by indirectly asking the student his values, satisfactions, appreciations, attitudes, etc., about what is being learned.

Shown below is a chart outlining these two types of behavior and some methods of determining their attainment:

TYPES OF BEHAVIOR	MEASURED BY
1. Overt	
a. Verbal	a. General observation
b. Non Verbal	b. Observational rating scales scored by the teacher

TYPES OF BEHAVIOR	MEASURED BY
2. Covert	
a. Attitudes	a. Self-evaluation inventories; support scales
b. Feelings	b. Likert scales; semantic differential scales; conversations
c. Values	c. Same as b.

Examples of these different types of measuring instruments are included in chapter eight.

Outlined on page 54 is a chart (3) showing the relationship of the overt and covert behaviors of the affective domain. Action verbs are stated for the verbal and non-verbal parts of the overt area.

Remember, the covert types of affective attainment are more difficult to determine. Usually the instructor will have to ask a student how he feels, values, appreciates, or he will have to construct instruments to measure these feelings. Self-evaluational inventories are especially useful for this purpose and several examples of these and the justification of their use are given in chapter seven. Brief examples are listed below to help you see their use.

Self Evaluation Inventory

1. I like mathematics better than science.
   Strongly agree Agree Disagree
2. Mathematics will probably not be very valuable to me in my life.
   Strongly agree Agree Disagree

Use of Scale For Science

Place an X in the space of your choice

1. Good ____ ____ ____ ____ ____ Bad
2. Fun ____ ____ ____ ____ ____ Dull
3. Sometimes pleasurable ____ ____ ____ ____ ____
4. Successful ____ ____ ____ ____ ____ Unsuccessful
5. Like ____ ____ ____ ____ ____ Dislike

Overt Aspects of Affective Achievements

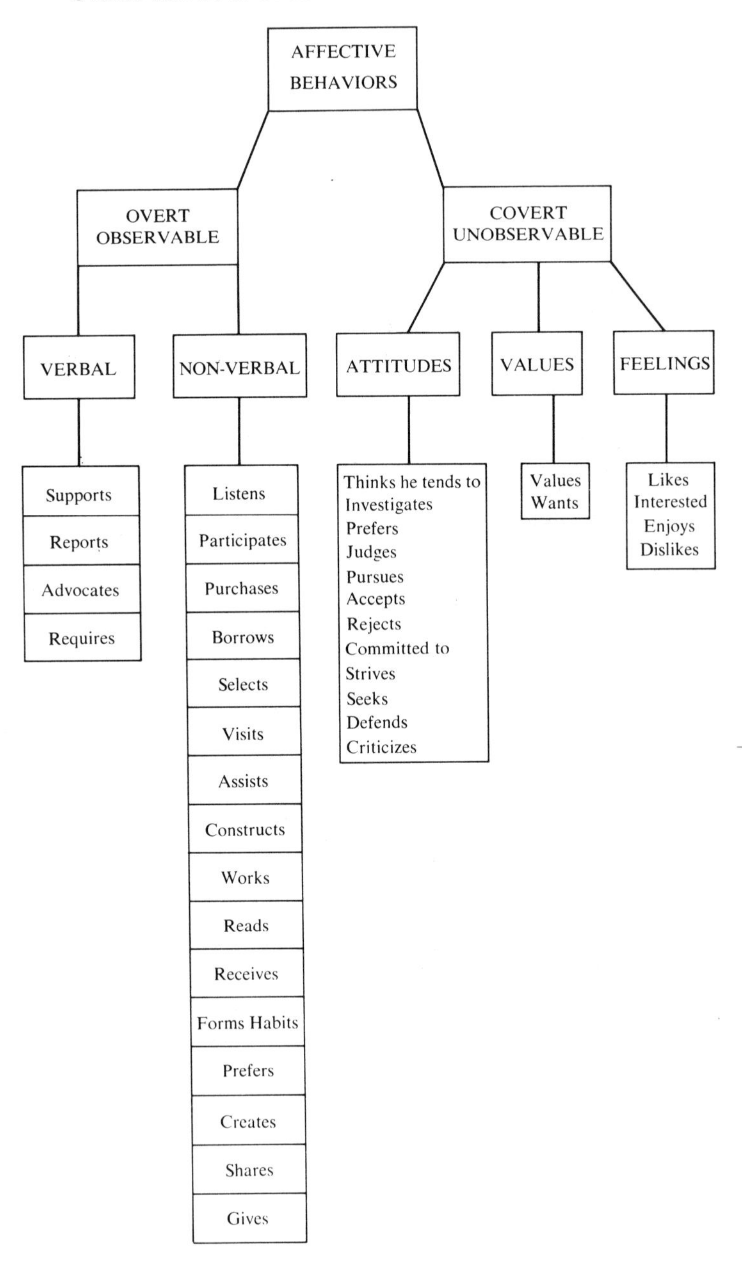

Sample Objectives

The following are general examples of affective objectives. They have been categorized as being "observable" or "unobservable" and subdivided into problem solving, general, and specific subject area objectives.

Observable Objectives

Problem Solving:

The student should:

1. *Volunteer* to do additional experiments in areas related to the subjects studied in class.
2. *Volunteer* to do additional work in the form of a written paper or oral presentation in areas related to subjects studied in class.
3. *Participate eagerly* in laboratory exercises.

General Objectives:

The student should:

1. *Join* (voluntarily) the science club.
2. *Volunteer* to attend science fairs that he has the opportunity to attend.
3. *Volunteer* to attend optional field trips to museums, universities, etc.
4. *Participate eagerly* in classroom activities.
5. *Offer* constructive criticism on the improvement of classroom activities.
6. *Complete* and hand in assignments on time.

Specific Objectives:

The student should:

1. *Initiate* group action dealing with pollution problems in our community, such as, organizing a working committee to determine and publicize sources of pollution.
2. *Promote* or *take part* in group and individual activities in preserving natural and living resources in the community.

Unobservable Objectives

Problem Solving:

The student should rate positively, on a self-evaluation inventory, on each of the following:

1. *Desire* to do additional work on biological topics or related topics not specifically studied in class.
2. *Consider* problems encountered in an objective manner. (Example — form conclusions based on logical thought processes, rather than on bias or emotions.)
3. *Apply* good "scientific technique" in laboratory work.
4. *Exhibit* self-confidence in terms of solving problems encountered in the classroom.

General Objectives:

The student should rate positively, on a self-evaluation inventory, on each of the following:

1. *Want* to take another biology course.
2. *Like* biology as much at the end of the year as at the beginning of the year.
3. *Recognize* the importance of, and *support*, scientific research.
4. *Voluntarily* read outside material related to biology.
5. *Enjoy* participating in group activities in class.
6. *Form* or *change opinions* on the basis of new evidence which is discovered.
7. *Respond* to suggestions and attitudes of classmates.
8. *Freely express* own opinion concerning topics or problems encountered in class.
9. *Be receptive* to new or different learning activities in class.

Specific Objectives:

The student should rate positively, on a self-evaluation inventory, on each of the following:

1. *Desire* to obtain additional information concerning the problem of world over-population.
2. *Make decisions* concerning the individual's responsibility for conserving natural and living resources.

3. *Make decisions* concerning society's responsibility for conserving natural and living resources.
4. *Make decisions* concerning the utilization of insecticides, herbicides, and other potentially dangerous substances as each relates to food production, disease control, etc.
5. *Utilize* good hygenic habits in every day life. (Example — maintain a proper diet, exercise regularly, etc.)
6. *Decide* not to begin smoking.
7. *Decide* to quit smoking if he has started.
8. *Make decisions* concerning the harmful effects of drugs and drug abuse.
9. *Become convinced* of the harmful effects of excessive amounts of alcohol and decide not to use alcohol in excess.

Continuums of Affective Behaviors

Krathwohl points out, in comparing the cognitive and affective domains, ". . . there is a tendency for the counterpart of a low level objective to come from the lower levels of the affective continuum and for objectives at the upper level of the affective continuum to have upper level cognitive counterparts (2, p. 53)." Not only is there a correlation between emotion and cognition, but there seems to be a positive correlation between the intensity of these behaviors. *The more a student "knows" about science and mathematics, the higher he values them. Conversely, a student who values science and mathematics will attempt to become more cognizant in these areas.*

The implication of the proceeding comments is that affective behaviors cannot be ignored. It cannot be assumed they will be achieved as a natural by-product of cognition. Specific objectives must be formulated relating to affective behavior in science and mathematics as carefully as those relating to cognitive behavior. There is admittedly more difficulty in stating observable affective behaviors than cognitive ones. There is also less precision in measuring the achievement of affective behaviors. Neither of these difficulties, however, is insurmountable. Kapfer suggests:

1. Stating the affective domain objective as an unobservable behavior (e.g., receiving, responding, valuing, organizing, characterizing) and then stating the related observable area of behavior.
2. Stating finite linear steps in a continuum of behaviors beginning at the "negative" or "neutral" end and progressing to the "positive" end of the continuum.

For example:

Objective: The student increasingly values independent learning, as observed in his self-initiating and self-directing behaviors.

Continuum of Behaviors

1. Given a teacher-assigned delimited topic with assigned specific resources, the student follows directions.
2. Given a teacher-assigned delimited topic and assigned alternative resources, the student selects from alternative resources.
3. Given a teacher-assigned delimited topic, the student seeks his own resources.
4. Given a teacher-assigned broad topic, the student delimits the topic and seeks his own resources.
5. Given a student-initiated broad or delimited topic (in or out of school), the student delimits the topic as necessary and seeks his own resources (1, p. 12).

How to Start to Write Affective Objectives

1. Determine your goals.
2. Initially concentrate on writing value-laden objectives. A beginning writer of affective objectives should ignore the first and second levels of *The Taxonomy*, those of receiving and responding, which are involved in many ways with describing the cognitive domain. To become concerned about writing objectives for the lowest levels of the hierarchy, and also for the fourth and fifth levels, tends to confuse the beginner as to the function of the affective domain. However, writing value objectives has the opposite effect. Furthermore, writing for valuing or higher objectives is more rewarding and has a greater effect upon the writer than stating them for the first two levels.
3. Write the following phrase at the top of the page once and only once:

   "The student should be able to:"
4. Select an action verb — refer to the examples of verbs used in the chapter and write your statement.

   The student should be able to:

   1. *Participate* in class discussion

2. *Like* mathematics as indicated on a self-evaluational inventory
3. *Enjoy* the class.

Notice that in one of the objectives above, the method for evaluating it is also included. It is probably better to write your objectives and then rewrite them including the method of evaluation and, if necessary, a hierarchy of the behaviors you expect.

Summary

The affective domain deals with appreciation, feelings, values, and attitudes involved in instruction. Affective behavior may be overt or covert. Overt behavior is evaluated by general observation or by the teacher using a scale as a guide to rate the student. Covert behavior is more difficult to determine since it is not immediately observable. Its achievement may be evaluated by asking the student how he feels about what he is learning, by using self-evaluational inventories, or by applying Likert or semantical differential scales. The steps and the form to follow in writing affective objectives are similar to those of the cognitive domain. The action verbs, however, vary, and it should be remembered that it is possible to describe a continuum of affective behaviors from those that are negative to positive.

It is suggested that a beginner start by writing valuing level affective objectives. Once he has become competent in writing these, he may want to construct objectives for the other levels of the affective hierarchy.

Bibliography

1. Kapfer, P. "Behavioral Objectives in the Cognitive and Affective Domains." *Educational Technology*, June 15, 1968, p. 12.

2. Krathwohl, David, Bloom, B., and Masia, B. *Taxonomy of Educational Objectives, The Classification of Educational Goals, Handbook II Affective Domain*. New York: David McKay Company, Inc., 1964.

3. Sharpe, Glyn H. "A Model for Assessing Affective Variables At the Curriculum Level." Unpublished paper, Colorado State Department of Education, 1970.

CHAPTER SIX

THE PLACE OF BEHAVIORAL OBJECTIVES IN CURRICULUM FORMATION

Organization of the Curriculum

There are several ways to construct a curriculum. One organization, followed by most elementary and secondary school districts, consists of a series of units. These are organized into a scope and sequence under one of several conceptual schemes (4, p. 20). An example in science would be, "Matter exists in the form of units which can be classified into hierarchies of organized levels." A second approach to curriculum building is to organize it into a hierarchy of mental activities. The main purpose of doing this is to engender the development of such mental processes as observing, classifying, inferring, controlling variables, etc. This curriculum design is based on the assumption that facts learned by the child may be obsolete in a few years. It is, therefore, more important that he *learn how to learn*. The American Association for the Advancement of Science in their *Science — A Process Approach*, has used this design in their materials. The third general pattern for curriculum construction attempts to integrate the teaching of concepts and principles with mental process development. This integrative format differs from the AAAS *Process Approach* in that no hierarchy of processes is stipulated. The type of process being developed evolves out of the nature of the phenomena being studied. This integrative approach has been used by the Elementary Science Study, Earth Science Curriculum Project, Chemical Education Materials, Project Physics, Physical Science Study

Committee, Biological Sciences Curriculum Study, and School Mathematics Study Group; it has been suggested by the Cambridge Conference on School Mathematics, University of Illinois Committee on School Mathematics, and most of the recent elementary and secondary science and mathematics text series. A fourth approach is to construct a curriculum to develop the creative abilities of students in science and mathematics. No curriculum in science or mathematics, however, has been organized for this purpose at the present time.

Designing a Curriculum

The first step in designing a curriculum is to stipulate its goals. The goals should then be translated into behavioral terms, activities devised to help the student achieve these objectives, and finally evaluational techniques outlined to determine how well the objectives are achieved.

Although there has been resistance by some educators to stating behavioral objectives, recent interest in programmed instruction and computer assisted instruction has necessitated that they be so defined. A program for a teaching machine or a computer cannot be written without stating what is to be specifically learned. In other words, the input and the output of the program must be determined. The advent of the computer has opened tremendous possibilities for diagnosing the ills of an instructional system. For example, when children respond to a computerized program, the time and nature of the response for each objective is recorded. The curriculum creator can look at the nature of the responses, identify areas of difficulty, and modify the program accordingly. If, for example, the third objective of the curriculum is not being attained as determined by the output from the computer, the designer can modify the input and retest the program with another group of students the next day. By so doing, the curriculum can be continually altered and improved at a much faster rate than presently is the case with most curriculum projects. The use of a centralized computer with many terminals in several schools also makes it possible to field test the curriculum with a relatively diverse population before it is released for wide distribution. This approach to curriculum construction is presently being used in reading and mathematics in the Computer Assisted Instruction Project at Brentwood Elementary School in Palo Alto, California, and in the Intermediate Science Curriculum Study at Florida State University. Dr. Robert Glaser, Director of the Learning Research and Development Center at the University of Pittsburgh, has pointed out that another advantage of using the computer in curriculum construction is that it may provide "a basis for redefinition of objectives." He says:

> As the curriculum designer and teacher see what really is possible, they may see more clearly the kinds of instruction and performance capabilities that students need in addition to those initially considered in lesson planning. The process of clarifying goals, working toward them, appraising progress, reexamining the objectives, modifying the instructional procedures to achieve goals, and clarifying the objectives themselves in the light of experience and data should be a continuous process. This process, however, has little focus in the absence of initially specified objectives (2, p. 2).
>
> A related point is that regardless of the way subject matter is structured there is usually present some hierarchy of subobjectives. This hierarchy may be influenced by the properties of different tasks and by individual differences: Nevertheless, the absence of prerequisite competence in a sequence of instruction dooms many students to failure. Specification of objectives permits identification and diagnosis of entering competence as well as terminal performance in a course of study (2, p. 2).

The Individually Prescribed Instruction Elementary Science Curriculum Project, with which Dr. Glaser has been associated, is an example of a curriculum that has behaviorally defined its objectives. Samples of some of them appear below.

Individually Prescribed Instruction Elementary Science Curriculum

Level A

Simpson Unit

Behavioral Objectives

1a. Given a group of objects of four or more different colors, the student sorts the objects into sets on the basis of color. (F)

1b. Given two or more sets of objects sorted on the basis of color, the student states that the objects are sorted on the basis of color. (I)

2a. Given objects or pictures of objects that are cubical or spherical in shape, the student identifies those objects that are cubes and those objects that are spheres. (F)

2b. Given an object that is cubical or spherical in shape, the student states the name (cube or sphere) of the object. (F)

3a. Given objects that are rectangular prisms or cylinders, the student identifies those objects that are rectangular prisms and those that are cylinders. (F)
3b. Given an object that is a rectangular prism or a cylinder, the student states the name (cylinder or rectangular prism) of the object. (F)

4a. Given objects that are cubes, spheres, rectangular prisms, and cylinders, the student identifies those that are cubes, those that are spheres, those that are rectangular prisms, and those that are cylinders. (F)
4b. Given a group of objects that are cubes, spheres, rectangular prisms, and cylinders, the student sorts the objects into sets on the basis of shape. (F)

5a. Given objects or pictures of objects that are square, circular, or triangular in shape, the student identifies those objects that are squares, those that are circles, and those that are triangles. (I)
5b. Given an object that is square or circular or triangular in shape, the student states the name (square or circle or triangle) of the object. (F)

6a. Given objects or pictures of objects that are elliptical or rectangular in shape, the student identifies those objects that are ellipses and those that are rectangles. (I)
6b. Given a group of objects that are square, circular, triangular, elliptical, or rectangular in shape, the student identifies those that are squares, those that are circles, those that are triangles, those that are rectangles, and those that are ellipses. (F)
6c. Given an object that is elliptical or rectangular in shape, the student states the name (ellipse or rectangle) of the object. (I)

7a. Given a group of objects of any shape, the student sorts the objects on the basis of shape. (F)
7b. Given two or more sets of objects sorted on the basis of shape, the student states that the objects are sorted on the basis of shape. (I)
7c. Given two or more sets of objects sorted on the basis of either color or shape, the student selects those sets which were sorted on the basis of shape *and* those sets which were sorted on the basis of color. (F)

8a. Given three different-sized objects of the same shape and of either the same color or different colors, the student identifies the largest, the smallest and the medium-sized object. (I)
8b. Given three different-sized objects of the same shape and of either the same color or different colors, the student describes the objects using the terms "largest," "smallest," and "medium-sized." (I)

9a. Given nine or more objects including three or more groups of three objects of the same shape in three sizes and of different colors, the student sorts the objects on the basis of size. (F)
9b. Given two or more sets of objects sorted on the basis of size, the student states that the objects are sorted on the basis of size. (F)
9c. Given two or more sets of objects sorted on the basis of either color or shape or size, the student selects those sets which were sorted on the basis of color, those which were sorted on the basis of shape, and those which were sorted on the basis of size. (F)

10a. Given four objects of the same shape, differing in height and width, the student identifies the tallest and the shortest. (I)
10b. Given four objects of the same shape, differing in height, and width, the student describes the tallest and shortest objects using the terms, "tallest" and "shortest." (F)
10c. Given four objects of the same shape, differing in height and width, the student orders the objects by height from tallest to shortest or shortest to tallest. (F)

11a. Given four objects of the same shape, differing in length and width, the student identifies the longest and the shortest. (I)
11b. Given four objects of the same shape, differing in length and width, the student describes the longest and shortest objects using the terms, "longest" and "shortest." (F)
11c. Given four objects of the same shape, differing in length and width, the student orders the objects by length from longest to shortest or shortest to longest. (F)

12a. Given four objects of the same shape, differing in width and length, the student identifies the widest and narrowest. (F)
12b. Given four objects of the same shape, differing in width and length, the student describes the widest and narrowest using the terms "widest" and "narrowest." (I)
12c. Given four objects of the same shape, differing in width and length, the student orders the objects by width from widest to narrowest or narrowest to widest. (F)

Level D

Lavoisier Unit Objectives
Scientific Literacy

Lavoisier IL–1

Given an unlabeled drawing of a burning candle that has the locations of various parts indicated by arrows and given a list of the parts of a burning candle, the child matches the indicated locations for all the following parts: solid wax, liquid wax, gaseous wax, unburned wick, burned wick, glowing wick, blue part of flame, clear part of flame, yellow part of flame.

Lavoisier IL–2

a. When asked what produces the changes in physical state of the wax in a burning candle from solid wax to liquid wax and from liquid wax to gaseous wax, the child states or writes that it is heat.

b. Given a list of the three physical states of wax and instructions to select the state of the burning wax in a candle, the child selects "gaseous wax."

c. Given a list of various kinds of energy and instructions to select from the list two products resulting from the burning of wax in a candle, the child selects both "heat" and "light."

Lavoisier IL–3 (filmstrip lesson)

a. When asked to give an explanation for the title of this unit about burning (or to select the best explanation from among three or more suggested explanations), the child states (or selects the one explanation which states) that Lavoisier was a scientist who studied burning.

b. When asked to tell about the life and work of Antoine Lavoisier, the child cites (or selects from a suggested list) at least three events or contributions which are presented in the Lavoisier filmstrip.

Lavoisier IL–4

a. Given a list of various substances and instructions to select from the list the one substance which is needed for wax to burn, the child selects "oxygen."

b. Given a picture of two closed jars of equal capacity, one jar containing an unburned candle and the other jar containing a smoking candle, and given instructions to select the jar with the larger quantity of oxygen, the child selects the jar containing the unburned candle.

c. When asked to give an explanation for why a candle burning in a closed container goes out after a period of time (or to select the best explanation from among three or more suggested explanations), the child states (or selects the one explanation which states) both that some of the oxygen in the container is used up and that not enough is left for the wax to burn.

d. Given several pictures illustrating activities the child has done in this unit, the child selects the picture illustrating the activity which demonstrates that water is formed as a candle burns.

Lavoisier IL–5

a. Given a list of various substances and instructions to select from the list those substances which are formed when wax burns, the child selects both "carbon dioxide" and "water."

b. Given several pictures illustrating activities the child has done in this unit, the child selects the picture illustrating the activity which demonstrates that carbon dioxide is formed as a candle burns.

Lavoisier IL–6

a. Given a list of suggested properties and instructions to select from the list two properties of all kinds of matter, the child selects both of these properties: "has mass" and "takes up space."

b. Given a list of suggested properties and instructions to select from the list two properties of energy, the child selects both of these properties: "does not have mass" and "does not take up space."

c. Given a list of various substances and various kinds of energy and given instructions to select all the listed kinds of energy, the child selects as many of the following that are on the list: heat, light, sound.

d. Given three or more suggested definitions and instructions to select the one best definition of the word "fuel," the child selects the definition: "a substance with stored-up energy."

e. Given a list of various substances and instructions to select all the listed substances which are fuels, the child selects as many of the following that are on the list: gasoline, hay, sugar, wax, wood.

f. Given a list of various substances and instructions to select from the list the one substance which is needed for a fuel to release its energy, the child selects "oxygen."

g. Given a list of various substances and instructions to select from the list those substances which are formed when a fuel releases its energy, the child selects both "carbon dioxide" and "water."

h. When asked to give an explanation of what happens when a substance burns (or to select the best explanation from among three or more suggested explanations), the child states (or selects the one suggested explanation which states) that when a substance burns it combines with oxygen and it releases energy.

Procedures For Curriculum Building

Outlined below are a list of steps to follow in writing a behaviorally oriented curriculum.

1. Determine the criteria for selecting objectives. The basis of any set of objectives depends on the philosophy of the project. The establishment of the philosophy is a prerequisite to the rest of the steps in curriculum construction. Often when school district personnel are involved in curriculum construction, considerable fric-

tion occurs because of the vagueness of what they are trying to do. Needless waste of time could be eliminated if criteria were established as a first step.

2. Determine the aims and then define them in behavioral terms. The goals or general objectives should be defined first. These should include cognitive, affective, psychomotor, and performance goals. Once these have been outlined, the specific or subobjectives should be determined and stated behaviorally for each domain. In the cognitive domain these will mainly be limited to statements involving the learning of science and mathematics concepts and principles.
3. Determine the hierarchy of objectives. In science and mathematics certain concepts and mental processes are essential to the learning of other concepts. These, therefore, must be ordered accordingly.
4. Outline the activities used to implement the objectives. Activities should be prepared to contribute to the attainment of both the general and specific objectives involving all of the domains.
5. Determine evaluation techniques to be used and prepare those needed for all the objectives. To most teachers, evaluation means testing. However, in science and mathematics, many evaluation techniques which do not require written tests are available to the teacher. For example, the best way to determine whether elementary children know that a magnet attracts or repels a substance is to give them magnets and see how they use them to attract or repel objects. Operationally, a magnet is what a magnet does. If children demonstrate that they know how to attract or repel an object with a magnet, they have achieved the objective.

   In mathematics, for example, the proof of the understanding of the metric system, is how children use it to measure phenomena. A teacher doesn't have to give a paper-pencil test to find this out.

   A behaviorally defined curriculum means that behaviors are observable. The evaluative techniques used in this type of curriculum are therefore more likely to be practical laboratory activities for the purpose of evaluation than has been the case with traditional, subject-matter-centered courses. This does not indicate that paper-pencil tests do not have a place. They do, but the items used on these tests are far more sophisticated since they are designed to test for all of the educational domains. A considerable amount of information is devoted to the construction of evaluative instruments in chapter nine on evaluation.

6. Define the competency levels. How much must a student know about an objective to have satisfactorily attained it? The answer to this question is particularly important if a curriculum is to be individualized. How will a teacher know when a student is ready to progress to the next unit of study? The instructor must have defined the minimal level of competency a student will have to attain to indicate he has achieved an objective.

   The determination of competency levels is also important in determining the success of the curriculum. It is of little value to have goals which are unattainable for a certain grade level or age group. If an objective is not being attained by 90 per cent of the students, it should be modified or rejected.

   The setting of competency levels also becomes particularly important where the subject matter progresses in difficulty and understanding, as is the case with geometry and certain areas of chemistry and physics. If students haven't mastered fundamental concepts, they certainly should not be allowed to progress to work of a more inclusive nature which is dependent upon mastery of the fundamentals.
7. Analyze the results of the evaluations, modify the curriculum materials as necessary, and retest. Continuing this process will eliminate poor areas of the curriculum and will insure a program which is up to date.
8. Redefine the objectives where necessary.

Science — A Process Approach Curriculum Organization

The American Association for the Advancement of Science has designed its *Science — A Process Approach* somewhat in the manner as stated above. It has stated its rationale for establishing, constructing, and evaluating the attainment of behavioral objectives as follows.

Behavioral Description of Objectives

1. If a curriculum claims to be helping the learner acquire certain behaviors, then there is an inescapable obligation on the part of the curriculum designers to specify the objectives of the curriculum. The description of the objectives must be specific, reliably observable behaviors.
2. If the objectives of the curriculum are described in terms of reliably observable behavior, then the curriculum developer must

assume the responsibility of constructing measures to assess whether the learner has acquired the behaviors described as the objectives.

3. If the instructional materials designed are to be useful, then there is an obligation on the part of the curriculum designer to demonstrate evidence of observable learner accomplishments (acquired behaviors) in terms of the specified behavioral objectives.
4. The behavioral objectives must be clearly and explicitly stated in writing.
5. The evidence of the presence of any described behavior must be obtained in such a manner that the procedure used to collect the evidence can be replicated by other independent investigators.

Guidelines
In the construction and assessment of behaviors described as objectives:

1. What performance is the learner expected to exhibit? (Action verbs serve this purpose.)
2. What situation initiates the learner's performance?
3. What objects are acted upon?
4. What constitutes an acceptable response?
5. What restrictions are imposed on an acceptable response?

Sources of Data
The following are the "four sources of data collected in the evaluation of this curriculum:"

1. The behaviors exhibited by children after each exercise.
2. The teacher's description of the preparation and execution of instruction.
3. The progress of the children in reaching the terminal behaviors.
4. The characteristics of the tryout teachers and centers (1, p. 3).

Competency Measure

The *Process Approach* curriculum project has also established a set of measures called *Competency Measures* to determine how children succeed in the attainment of their objectives.

> The competency measure tasks are designed to assess each behavior described as an objective of the exercise. An additional characteristic of the competency measure tasks is that they are purposely designed to involve changes of stimulus or context from the instructional activities of the exercise. The competency measures provide data on whether the child has acquired the behaviors described as objectives. They do not test rote recall (1, p. 3).

The following are examples of the behavioral objectives and competency measures of an exercise, *Density, Defining Operationally 6, Part Seven.*

Examples of Science Process Approach Objectives

1. Demonstrate a procedure for finding the volume of a solid object by measuring the volume of water it displaces.
2. Construct an operational definition of density.
3. Demonstrate the computation of the density of an object given the volume and mass of the object.
4. Distinguish between objects which will float in water on the basis of their density (1).

Evaluation of the Objectives (Competency Measures)

1. The teacher gives a child six BB's and asks what is the volume of these six BB's?
2. What is the mass of these six BB's (1)?

The teacher has a book containing competency measures for each exercise. The questions to be asked, as well as any other instructions to be given by the tester, are identified in the competency measures. The teacher observes what the children do with the BB's and records their competency in solving the problem. The measures, therefore, do not have to rely on paper-pencil tests, but rather on learner performance and teacher observation.

How does the AAAS know that their exercises are successful? Their aim is to have 90 per cent of the children attain 90 per cent of the desired behaviors. It is through the evaluation of the competency measures that the AAAS determines whether a particular activity should be accepted or rejected.

Science Process Instrument (SPI)

As with any curriculum project there is a problem of introducing sequential materials so the children beginning to use them are properly challenged. There is also the problem of placing a child who transfers into the school and introducing him to the materials at his level of competency. For this reason the AAAS Project has constructed the *Science Process Instrument.* This consists of separate measures for each of the objectives outlined in the project. Each of the behaviors described in the hierarchy of the processes being developed is represented by a task on the process measure. The AAAS outlines the functions of the SPI as follows:

1. A method of evaluating the level at which to begin instruction with a group that has not previously been involved in *Science — A Process Approach.*
2. A means of identifying individuals who share common difficulties or who have acquired similar behaviors so that they can be grouped together for instruction, either in a team-teaching situation or for small group instruction.

3. A diagnostic tool which can be used when individualizing instruction within a class. The behaviors which specific individuals have not acquired can be readily identified and activities from the program can be used to develop those behaviors.
4. A regularly administered measure to determine acquired behaviors. When the instrument is being used in this capacity, it is anticipated that it will be administered at the beginning or the end of the school year or possibly on both occasions. The information gained from this testing provides a description of the child which can be entered in his permanent record.
5. A basis for reporting to the parents the achievement of their children (1).

Other instruments the AAAS has developed are the *Extra School Information Inventory* for testing interests and attitudes and *Teaching Time Inventory* to determine the average time used by each teacher.

Several other national projects have followed a plan of operation similar to the AAAS in designing a new curriculum. The behaviorally oriented curriculum is a breakthrough in the complex problem of curriculum construction. Because of the diagnostic value the statement of behavioral objectives allows in determining curriculum weaknesses, strengths, and placement of students, this approach in building and refining a curriculum will become more prevalent. The compatibility of this technique with computer methods will enable some of the tedious components to be performed by the machine. The pattern outlined and implemented by the AAAS will, undoubtedly, become a guide for similar projects.

Summary

There are four fundamental ways to design a curriculum. They are: the conceptual scheme, process-centered approach, integrative (process and conceptual) scheme, and creative approach. National projects have constructed curriculums around the first three of these. No curriculum, however, has been written specifically to develop creative ability. Science and mathematics texts have largely followed the conceptual scheme or integrative approaches.

Many of the new science and mathematics curriculum projects have defined their objectives in behavioral terms. These include the Individually Prescribed Instruction Project, *Science — A Process Approach*, plus the various computer assisted projects. *Science — A Process Approach* is a model of how the behaviorally designed curriculum is constructed and operates. This project establishes criteria for selection of their objectives, defines objectives behaviorally, outlines competency measures for each objective, uses *Science — A Process Instrument* to determine where to place a student in the curriculum, has built an *Extra School Information Inventory* for listing interests and attitudes, and has constructed a *Teaching Time Inventory* to determine the average time used by a teacher to teach each of the lessons.

The behaviorally defined curriculum has the advantage of being relatively easily evaluated and more appropriately modified than a non-behaviorally oriented curriculum. The steps to be taken in writing a behavioral curriculum are:

1. Determine the criteria for selecting the objectives.
2. Determine its goals and then define them in behavioral terms.
3. Determine the hierarchy of its objectives.
4. Outline the activities to implement its objectives.
5. Determine and prepare evaluational instruments.
6. Define the competency levels for each objective.
7. Field test and evaluate the curriculum, then modify it where needed.
8. Redefine objectives where necessary.

A behaviorally defined pattern of curriculum design is likely to be the mode for many new projects.

Bibliography

1. Commission on Science Education. *News Letter*, American Association for the Advancement of Science, vol. 3, no. 3, June 1967.

2. Glaser, Robert, "Objectives and Evaluation: An Individualized System." *Science Education News*, June 1967, p. 2.

3. Lipson, Joseph, et. al. "Individually Prescribed Instruction." *Elementary Science Curriculum*, Learning Research and Development Center, University of Pittsburgh, 1967.

4. National Science Teachers Association Curriculum Committee. *Theory Into Action*. National Science Teachers Association, Washington, D.C., 1964, p. 20.

CHAPTER SEVEN

THE UNIFICATION OF SCIENCE AND MATHEMATICS

The Drive for Integrated Science and Mathematics

Science and mathematics educators have been concerned about the isolation of science from mathematics in the school curriculum. This concern mounted to such an extent that a special group of leading scientists, mathematicians, and educators convened during the summer of 1967 in Massachusetts. The members of this group, which was called the Cambridge Conference, generally agreed that the advantages of integrating science and mathematics far outweighed the disadvantages. A report of the Conference stated:

> An integrated math-science curriculum will be difficult to implement, and perhaps the greatest difficulty will be the problem of training teachers to handle the material. Nevertheless, it appears that an integrated curriculum designed to bring out these connections is necessary in view of two facts: quantitative thinking is the essence of the power of the scientific method; many pupils are unable to grasp the connections between the mathematics they are taught and the real world. This curriculum would comprise a variety of units and activities which could be variously described as:
>
> ________ math for math
> ________ math for science
> ________ math *and* science
> ________ science for science (3)

This indicates that the members of the conference were of the opinion that there would still have to be some attention given to special topics of mathematics and science, but a major portion of those disciplines could be integrated. The group realized an integrated curriculum could not be developed quickly since years would be required for the needed experimentation. The following are some suggestions they made for such a curriculum:

1. The units should be open-ended allowing for further investigation by students motivated by the topic.
2. There should be a natural mixture of exploratory observations.
3. Attention should be given to measurement in the elementary school.
4. Functions, since they are frequently used models of reality, are "central to math-science correlation." They should be developed in the early grades, and children should learn to prepare graphs illustrating these functions. (It was suggested that such involvement would also better contribute to an understanding of ratios.)
5. More time should be devoted to learning and using estimation on slide rules and calculators.
6. Since the world is not two-dimensional, but three dimensional, studies involving three dimensional objects, such as, plyhedra, spirals, snailshells, and highway clover leafs should be incorporated into the children's studies.
7. The possible use of computers should be considered, especially where a student is given an opportunity to program, because this requires him to organize his ideas carefully (3).

The conference participants criticized the present science curriculums as being mainly qualitative. An elementary child, for example, may learn that fertilizer, when added to plants, stimulates their growth, but he may not learn how much fertilizer is used or desirable. Dr. Andrew Gleason, Chairman of the Conference, has pointed out that qualitative observations do not give the child the impression of the spirit of modern science, since science usually evolves by collecting numerical data and analyzing it (4).

Rainer Weiss of Massachusetts Institute of Technology, a member of the Conference, gave another reason for bringing science and mathematics together in the curriculum. He said:

> It may be that the academic flavor of both disciplines compartmentalizes thinking about math and science, so that the cognitive style — which we call scientific method — is not carried over into

life . . . Clearly a necessary attribute if the educational process is to arm the civilian against the expert (6, p. 16).

Similarity of Science and Mathematics Objectives

Modern science and mathematics educators have become increasingly concerned about students' levels of cognition during instruction. The word "cognition" comes from the Latin word *cognoscere* meaning to come to know and is used by behavioral scientists to refer to thinking processes. What processes are involved in the scientific method and how do they overlap with mathematics? A group of teachers involved in the Cubberley Lockheed Science Project in Palo Alto, California, have analyzed scientific problem solving, and they have defined the thinking behaviors of students who are involved in the act of being scientific investigators. This list appears below.

THINKING TACTICS

1.0 Indicate the possible affect of various qualifications or conditions on validity of the hypothetical answer.
- 1.1 State a problem to be solved.
- 1.2 Formulate hypothetical answers to the problem.

2.0 Recall previously learned processes, directions, and movements of phenomena with respect to time to find situations appropriate and analogous to a problem being studied.
- 2.1 State a problem to be solved.
- 2.2 Formulate hypothetical answers to the problem.
- 2.3 Recall methods or techniques used to make historical discoveries.
- 2.4 Recall observations which led to a generalization.

3.0 Use a generalization and its associated observations to make predictions about the nature of future observations.
- 3.1 Recall observations which led to a generalization.
- 3.2 Make a prediction of the results of given experimental procedure.

4.0 Relate various phenomena (structures, processes, theories).
- 4.1 Describe observed phenomena in an objective, non-teleological, non-anthropomorphic manner.
- 4.2 Recognize similarities and dissimilarities among two or more structures, systems, processes or theories.

5.0 Relate operational definitions to conceptual (theoretical definitions) and write operational definitions.

6.0 Make predictions based on interpolations and extrapolations of ordered data.

   6.1 Plot the tabulated data of two variables.
   6.2 Interpolate and/or extrapolate from the plotted data.

7.0 Recognize similarities and dissimilarities of parts and structures of compared organisms.

   7.1 Draw observed organisms such that relevant characteristics are shown.
   7.2 Label parts and structures shown in diagrams.

8.0 Judge the validity of a conclusion.

   8.1 Recognize or identify unstated assumptions used in a reported experiment.
   8.2 Differentiate between dependent and independent variables in the same reported experiment.
   8.3 Differentiate between the experimental and control conditions in the same reported experiment.
   8.4 Relate the conclusion of the reported experiment to the assumptions, variables, and conditions as described in the report.

9.0 Distinguish the conclusion based on observations from associated statements that tend to support it or stem from it.

   9.1 Recall observations that led to a generalization.
   9.2 Distinguish empirical observations from value judgments that are associated with them.
   9.3 Distinguish a conclusion from statements which support it.

10.0 Use the criteria and the attributes to analyze, synthesize, and criticize the terms, concepts, principles, and opinions (this one could be separated into at least three equivalent tactics).

   10.1 Recall criteria used to judge certain facts, principles, opinions, and experimental results.
   10.2 Recognize the attributes, properties, and relations of technical terms, concepts and principles (8).

It should be apparent to you, from your analysis, that most of these general objectives are relevant to mathematics as well as science.

This list could be used as a basis for devising activities involving science and mathematics problem-solving tactics. Since many science and mathematics courses attempt to provide for the development of these mental skills, it would seem to be more economical to develop them in an integrated science-mathematics framework.

Quantifying Reveals Patterns in Nature

Scientists strive to find patterns in nature. In so doing, quantification is often involved in collecting data and interpreting from it. Charles H. D'Augustine says of a student in this respect:

> . . . In his quest for these patterns he often lists his quantitative data in a table or translates it into a graph. After he has collected and organized his data, he attempts to construct a generalized mathematical model that will facilitate predictions about future events. He then tests the reasonableness of his generalized model by making further observations (2, p. 645).

D'Augustine goes on to give three examples of how mathematics may be involved in the conceptualizing of patterns. These are:

1. *Linear Relationship*

   His first example involves a child studying various stresses (weights) applied to a wire and the measurement of its consequent elongation. The child notes how the wire lengthens after each weight is added, goes on to graph the data and discover it gives a straight line graph (a linear relation).
2. *Statistics and Logic*

   The second type of problem involves the determination of the characteristics of offspring when short and long legged water bugs are interbred or crossbred.

   In analyzing and collecting data to answer these problems, the students are involved in making predictions. After completing several experiments, they finally develop a genetic model.
3. *Insight into the Dynamic Nature of Problem Solving*

   The third type of pattern searching involves a scientist interested in predicting the temperature of an area. He collects temperature data from various parts of the U. S. He soon discovers the recorded temperature in close proximity, at the same elevation as the area he is studying, does not change abruptly. He also discovers that by knowing the temperatures of four points around the area and averaging them, he can predict its temperature. This number will be relatively close to the temperature of

> his area. Fronts, however, change the picture somewhat. As he investigates more and more, he finds many factors that effect his predictions. His goal is to continue refining his "mathematical model " (2, p. 648).

D'Augustine points out that although these math skills are not inclusive, they give an indication of the types of integrative activities that should be developed in the elementary school. Furthermore, this approach of combining science and mathematics in an effort to give students better understanding of how nature's patterns are determined has the added advantage of being exciting to children.

N. F. Newbury has found that the learning of science in an integrative manner is fun for children even in the primary grades if their studies involve things in their daily experiences. He suggests:

> As an essential part of their training the children should individually guess the result before the activity is carried out. The checking of the answer by some precise form of measurement and use of suitable units enables a child to find out how accurate his guess or estimation was. Using worthwhile problems, appropriate ways of measuring the degree of accuracy should gradually be introduced (7, p. 641).

Some of the examples he gives that lend themselves to this purpose are:

> How is the pulse rate measured? Does it always beat at a steady rate? Could it be used instead of a watch for carrying out time measurements? Does the pulse beat at the same rate for all children of the same age? Is the rate the same before as after vigorous activity or when one is calm or emotionally upset? What is the pulse rate of a dog, a cat, a bird? How many pulses are there in the body? The children themselves will supply other questions, and determining the result involves much mathematics such as averages, minima, maxima, and graph (7, p. 641).

Other examples that he gives in various places in his article are:

> How much do I weigh in water?
>
> How far can you see on a clear day?
>
> What is the human hair frequency?
>
> The length of arms and legs and height of body.
>
> Find out the area of the feet and weight of a human body on a small area.

> What is the human body's area?
> How much air can be breathed out?
> How many bricks and how much cement are there in a wall? (7)

The Contributions of the Discovery or Inquiry Approaches to the Integration of Science and Mathematics

The integration of science and mathematics has undoubtably received considerable impetus from the stress modern curricular reforms in both of these disciplines place on investigation. This orientation has become known as the discovery or inquiry approach. The rationale for accepting this approach is that it develops critical thinking, problem-solving abilities, and the creative potential of students better than the traditional, fact oriented curriculum. Edith Biggs, speaking of a curriculum project with which she has been associated, typifies this viewpoint when she says:

> Above all, we provide opportunities for the children to think for themselves, so that learning for them is an active, creative process. Our aims in the teaching of mathematics at all levels are, in summary, to give our students 1) the opportunity to think for themselves, 2) the opportunity to appreciate the order and pattern which is the essence of mathematics, not only in the man-made world, but in the natural world as well, and 3) the needed skills (1, p. 405-06).

Work is continuing on devising better discovery- or inquiry-oriented curriculum projects. Since science and mathematics educators are both interested in developing creative problem-solving behaviors, greater integration should be apparent in future curriculum modifications.

Efforts to Integrate Science and Mathematics by Curriculum Projects

There have been several curriculum projects making efforts to integrate science and mathematics. Many of these have incorporated materials similar to those suggested by D'Augustine and Newbury. One project, MINNEMAST, was established specifically for this purpose in 1961 at the University of Minnesota. The initial efforts of MINNEMAST were devoted to developing mathematics-science units for grades K–9. This project has produced teacher guides, pupil work pages, and stories related to the lessons. The following units for grades K–3 have been prepared and field tested:

Kindergarten

Watching and Wondering
Describing and Classifying
Our Senses
Shape and Symmetry

Grade 1

Objects and Their Properties
Changing and Unchanging Properties
Introduction to Measurement
Time: Measurement of Duration
Locating Objects
Investigating Systems

Grade 2

Measuring Weight
Scaling and Representation
Time-Ordered Events
Observing Motion

Grade 3

Related Properties
Discovering Systems
Time Rates and Time Dependence
Investigating Systems in Stationary States

Another curriculum project interested in integrating mathematics and science is the American Association for the Advancement of Science, *Science — A Process Approach.* This project has produced K–6 materials. Its fundamental purpose is to develop such science process skills as: observing, classifying, measuring, using numbers, using spatial and temporal relations, making inferences, and predicting. Once these are learned, the child should be able to become involved in more complex skills, such as, interpreting data, formulating hypotheses, defining operationally, controlling variables, experimenting, and formulating models. John Mayor, Director of Education of the AAAS and a former president of the National Council of Teachers of Mathematics, has surveyed the AAAS units for those which include a considerable amount of mathematics. The results of his findings appear below:

TABLE 1

Exercise	*Title*	*Grade*	*Serial No. in part*
Using Numbers 1	Sets and Their Members	K	6
Using Numbers 2	Order Properties	K	12
Using Numbers 3	Numerals and Order	K	16
Using Numbers 4	Counting and Numerals	K	20

TABLE 1 (cont.)

Exercise	*Title*	*Grade*	*Serial No. in part*
Using Numbers 5	Numbers and the Number Line	1	6
Using Numbers 6	Numbers 0 through 99	1	11
Using Numbers 7	Addition of Positive Numbers	1	18
Using Numbers 8	Addition of Numbers Using the Number Line	2	3
Using Numbers 9	Multiplication	2	15
Using Numbers 10	Dividing to Find rates and Means	3	4
Using Numbers 11	Metersticks, Money, and Decimals	3	8
Using Numbers 12	Naming Large Numbers	3	23
Using Numbers 13	Decimals	4	5
Using Numbers 14	Decimals*	5	11
Using Numbers 15	Large Numbers, Glurks, and Respirations	5	16

TABLE 2

Exercise	*Title*	*Grade*	*Serial No. in part*
Measuring 1	Comparing Lengths	K	11
Using Space/Time Relationships 7	Symmetry	1	2
Measuring 2	Linear Measurement	1	4
Measuring 3	Comparing Volumes	1	10
Measuring 4	Linear Measurement Using Metric Units	1	12
Communication 2	Introduction to Graphing	1	16
Measuring 6	Ordering Plane Figures by Area	1	21
Communicating 7	Graphing Date	2	7

*As extension of Using Numbers 13

TABLE 2 (cont.)

Exercise	*Title*	*Grade*	*Serial No. in part*
Using Space Time Relationships 13	Straight Lines, Curved Lines, and Surfaces	2	11
Measuring 12	Measuring Volumes	2	21
Communicating 10	Maps	3	9
Communicating 11	Describing Location	3	10
Using Space/Time Relationships 15	Two-Dimensional Representations of Spatial Figures	3	19
Measuring 14	Measurement of Angles	4	6
Inferring 9	Inferring Shapes of Cut Things	4	13
Communication 12	Selecting Coordinate Systems for Graphs	4	15
Interpreting Data 5	Introduction to Probability	4	19
Interpreting Data 13	Contour Maps and Three-Dimensional Coordinate Systems	5	14
Measuring 16	Measuring Small Things	5	17
Formulating Hypotheses 3	Images of Objects	5	19
Interpreting Data 15	Probability by Experiment	5	20
Experimenting 5	Variation in Perceptual Judgments: Optical Illusions	6	6
Interpreting Data 16	Relations	6	8
Experimenting 17	Probability-Normal Distribution	6	17

(6, p. 863)

The *Study of a Quantitative Approach in Elementary Science*, completed in 1967, is a project having as one of its objectives that science be presented as a discipline usually requiring quantitative treatment. This project is being published by Scott, Foreman, and Company. Several hundred lessons, organized into units, have been produced, but no attempt has been made to build these into a spiral curriculum. Another curriculum project, the School Mathematics Study Group (SMSG), has also developed integrative science and mathematics units. One set of these units, entitled 'Mathematics through Science," involves three parts:

Measurement and Graphing

Graphing, Equations, and Linear Functions

An Experimental Approach to Functions

Another unit relating mathematics and biology is called "Mathematics of Living Things." Each of these curriculum projects includes a student text and a teachers supplementary text.

Many other curriculum projects have science lessons requiring the use of mathematics. These projects, however, have not been specifically designed to have children learn mathematics and science simultaneously, but mainly to show how mathematics is used in scientific investigation. Most of the modern commercially published science series have accepted a similar philosophical viewpoint. The amount of mathematics used in these new curriculum materials and text series, however, is considerably more than was the case previously. Furthermore, curriculum revisions to integrate science and mathematics are not limited to the United States. The Nuffield Project in Great Britain has done considerable work to bring about such an integration (1).

From this brief survey, it is apparent most of the efforts to integrate the learning of mathematics and science have mainly occurred in the elementary level. However, this is not to say there haven't been alterations of the junior and senior high school curriculums to include integrated mathematics and science materials. The junior high school level Introductory Physical Science and the Earth Science Curriculum Project both require a considerable amount of mathematics. The S.M.S.G. materials mentioned earlier were developed for junior high school students. On the high school level, efforts have been made, particularly in physics, to teach the mathematics needed to study certain physical phenomena. It seems reasonable that future integrative efforts will continue to be devoted to the elementary level, but once this task is underway, the secondary school curriculums will undoubtedly receive greater attention.

Summary

Science and mathematics educators have stressed the importance of integrating science and mathematics, particularly in the elementary school. The Cambridge Conference stated that the advantages of integrating these fields far outweighed the disadvantages, particularly since quantitative thinking is basic to the scientific method.

Many objectives of science and mathematics are similar, for example, the development of critical thinking. Science strives to discover patterns in nature through the use of critical thinking, and mathematics helps to reveal these patterns by showing relationships, using statistics and logic, and involving students in the dynamics of problem solving. Furthermore, the integration of science with mathematics is exciting to students, particularly if their investigations involve things from daily experience.

The discovery or inquiry approach used in modern curriculum projects has in many instances aided in integrating science and mathematics. Some of the projects having made considerable progress in doing this are: The *Science — A Process Approach*; the *Study of A Quantitative Approach in Elementary Science, The School Mathematics Study Group,* and MINNEMAST. Modern textbook series have made greater efforts to use mathematics where appropriate in their science units. The greater use of mathematics in the physical and biological sciences is likely to receive greater attention in curriculum design in the future.

Bibliography

1. Biggs, Edith. "Mathematics Laboratories and Teachers Centers — The Mathematics Revolution in Britain." *The Arithmetic Teacher*, vol. 15, no. 5, pp. 400-08, May 1968.

2. D'Augustine, Charles H. "Reflections on the Courtship of Mathematics and Science." *The Arithmetic Teacher*, pp. 645-49, December 1967.

3. Gleason, Andrew M., et al., "Simposium: Towards an Integrated Mathematics-Science Curriculum in Public Schools. Unpublished paper, American Assciation for the Advancement of Science, 134th meeting, December 29, 1967.

4. Gleason, Andrew M. "Science, Math, and Tomorrow's Child, Report from the Cambridge Conference." *The Instructor*, pp. 54-56, January 1968.

5. Gleason, Andrew M., et al. "Conference Focuses on Relationship of Math and Science in Education." MinneMast, Center Reports, vol. 5, no. 4, p. 16, 1968.

6. Mayor, John R. "Science and Mathematics in Elementary School." *The Arithmetic Teacher*, pp. 629-635, December 1967.

7. Newbury, N. F. "Quantitative Aspects of Science at the Primary Stage." *The Arithmetic Teacher*, pp. 641-44, December 1967.

8. Staff Members, Science Department Cubberley High School, Interim Report, Cubberley-Lockheed Science Project. "A Development Program to Attain Stated Behavioral Objectives in Science." Palo Alto Unified School District, Palo Alto, California, pp. 13-14, 1967.

CHAPTER EIGHT

THE USE OF SELF-EVALUATION-INVENTORIES

Self-Evaluational Inventories

Great efforts have been made in education to assess students' work through achievement tests. Seldom, however, have teachers asked students to evaluate themselves. This situation is changing because of the increasing development and use of self-evaluational inventories.

Self-evaluational inventories (S.E.I.) are instruments whereby students judge their progress toward the attainment of course objectives, evaluate the instructor, rate the types of activities included in a course or estimate their "self-esteem."

Students generally do not mind filling out self-evaluational inventories, especially if they think the instructor is using them as a guide for improvement. Allowing students to evaluate themselves on objectives has the advantage of increasing their awareness of the breadth of the objectives of the course. By taking the inventories several times, students become aware that they have learned and progressed.

Self-evaluational inventories can also be used by an instructor to diagnose the learning environment. By reviewing a student's rating of each objective on the inventory, the teacher gains information about the strengths and weaknesses of that student. By making similar reviews for all the students, the instructor easily sees objectives rated as high or low. By such an analysis, the teacher can identify those objectives requiring greater emphasis.

Research Indicates Students Evaluate Themselves Realistically

The works of Duel (1), John (5), Spicola (13), Lowther (7), Reince (11), Mowers (9), Fletcher (2), McCormack (8), and Tillery (13) indicate that students are able to rate themselves realistically. Tillery, for example, showed there was a high correlation between a student's rating on a self-evaluational inventory and his achievement on a test for each of the objectives of a course. It seems reasonable, therefore, that this type of instrument should be used more frequently than it has been. S.E.I.'s have the additional advantage of revealing how a student changes covertly.

Students' Feelings About a Subject Should Be Determined Prior to Instruction

The traditional view of teaching mainly stressed the importance of students' learning information. This outlook no longer is accepted by most educators. Rather, they believe it is the function of the instructor to help develop all of the talents — cognitive, psychomotor, and affective — of their students. How individuals feel about what is being learned has much to do with what they learn. For example, if a student doesn't think he can learn mathematics, he will not learn it as well as if he thought he could. Teachers should, therefore, endeavor to determine individuals' self-concepts and feelings about a subject area prior to the time they are involved in it. If a student has a negative feeling about the subject — a poor "self-concept"—the teacher should make special efforts to change this negative view to a positive one. This can be done by insuring the student has success in his work and by giving him positive reinforcement, i.e., "You are doing the chemical problems well." "I have noticed your work in the laboratory, and it is good." "You are a good mathematics problem solver," etc.

Because of the influence of a student's feelings about a subject on his learning, teachers should construct and administer self-evaluation instruments to determine a student's attitudes prior to instruction. The results from these assessments should then be evaluated by the teacher, and instruction modified accordingly.

Types of Self-Evaluational Inventories

Shown below are several self-evaluational type instruments. These are used to determine how a person *thinks of himself* relative to his general competence as a person (or his ability to do mathematics or science).

Remember, how a person perceives his ability to do something (his self-concept) is an excellent indication of *how* he will behave. A major role of a teacher is to change negative attitudes and build positive self-concepts.

How I See Myself In Science

This form allows you to express your feelings about science. There are several statements concerned with you and your work in this class. Read the statement and decide if you strongly agree, moderately agree, slightly agree, are neutral, slightly disagree, moderately disagree, or strongly disagree. Upon making your decision, circle the appropriate number. This will not be graded; it is only a guide to help me assist you.

Strongly Agree	*Moderately Agree*	*Slightly Agree*	*Neutral*	*Slightly Disagree*	*Moderately Disagree*	*Strongly Disagree*
1	2 3	4 5	6	7 8	9 10	11

Statement											
People like me.	1	2	3	4	5	6	7	8	9	10	11
I think I will succeed in this class.	1	2	3	4	5	6	7	8	9	10	11
I saw myself as successful in other science classes.	1	2	3	4	5	6	7	8	9	10	11
I view myself as an average student.	1	2	3	4	5	6	7	8	9	10	11
I view myself as an average citizen.	1	2	3	4	5	6	7	8	9	10	11
I see myself as popular.	1	2	3	4	5	6	7	8	9	10	11
I am a leader in the classroom.	1	2	3	4	5	6	7	8	9	10	11
I can effectively speak in front of a group of students.	1	2	3	4	5	6	7	8	9	10	11
I am apprehensive of this course.	1	2	3	4	5	6	7	8	9	10	11
I am adequate in my relations with other people.	1	2	3	4	5	6	7	8	9	10	11
I see myself as a reliable person.	1	2	3	4	5	6	7	8	9	10	11
I think others see me as reliable and dependable.	1	2	3	4	5	6	7	8	9	10	11
Students like me as a science partner.	1	2	3	4	5	6	7	8	9	10	11
Other students feel I contribute positively to class discussions.	1	2	3	4	5	6	7	8	9	10	11
Students in other classes like me as well as the students in this class.	1	2	3	4	5	6	7	8	9	10	11
People outside of school like me as well as students in this school.	1	2	3	4	5	6	7	8	9	10	11
I am about average concerning physical attraction by the opposite sex.	1	2	3	4	5	6	7	8	9	10	11
I am happy.	1	2	3	4	5	6	7	8	9	10	11
I am optimistic.	1	2	3	4	5	6	7	8	9	10	11

Student Attitudes toward Mathematics

According to Johnson and Rising, the five most recurring reasons students give for negative attitudes toward mathematics are:

1. Lack of understanding of mathematical principles
2. Lack of application of mathematics to a life situation
3. Too many boring problems assigned daily
4. Uninspired, impatient, uninteresting teachers
5. Lack of success (6, p. 130).

The following evaluation device might be used to assess these negative attitudes and to prevent them from being reinforced. This device is intended to be used at the conclusion of a unit of mathematical work.

How I See Myself in Mathematics

Answer each of the following questions by circling one of the responses below:

1. Definitely not
2. No
3. Can't decide
4. Yes
5. Definitely so

A. Do you understand most of the mathematics in the unit just covered? (List areas you didn't understand on reverse side.)	1 2 3 4 5
B. Do you see any way in which the mathematics might be applied to practical, real-life situations?	1 2 3 4 5
C. Were the daily assignments too long?	1 2 3 4 5
D. Were the daily assignments interesting?	1 2 3 4 5
E. Was the mathematics presented in an interesting way? (List ways in which it could be made more interesting on reverse side.)	1 2 3 4 5
F. Do you think the teacher expected too much of you? (List specific examples on reverse side.)	1 2 3 4 5
G. Would you say you were successful on this unit?	1 2 3 4 5
H. Would you study more about this unit if you had a chance?	1 2 3 4 5

Students Compare Their Cognitive and Affective Achievement With a Test of the Objectives of the Course

A technique to determine achievement or evaluate the teacher is to give students lists of objectives for a course and a rating scale. The students read each objective and use the scale to rate their achievement on them. Once the rating scale forms are prepared, they can be used as a general type of S.E.I. answer sheet. This kind of self-evaluational inventory is particularly helpful in evaluating the affective domain.

Listed below is a portion of the objectives of a course using this type of self-evaluation instrument for students. The objectives are listed first items.

Objectives for the Secondary Methods Course in Teaching Science

At the completion of this course you should be able to:

State the Nature of the Scientific Enterprise

A. Define terminology used in science, e.g., experiment, theory, and inductive logic, plus those passed out by the instructor on a glossary sheet.
B. Define and construct statements of problems, hypotheses, data, tests of hypotheses, and statements of conclusions.
C. Describe how a scientist goes about forming a theory, e.g., how he supports his ideas.
D. State examples of "scientific attitude."
E. Select, when given a group of activities, those helping to understand the nature of science (12).

Self-Evaluation Inventory Scale

In order to improve a course, it is helpful to know the amount of knowledge or skill each student possessed at the time he entered the course, compared to when he finished it. You are requested to indicate your opinion of the degree of skill or knowledge you possessed in each case for the following series of items.

Place two letters on the scale on the answer sheet for each item. Evaluate each statement in terms of the scale shown. Descriptions of various points on the scale appear above it. "B" should be used to indicate the amount of skill or knowledge you had at the *beginning* of the course, and "E" should be used to indicate the degree of skill or knowledge you now have, at the *end* of the course.

I have no knowledge of this objective.	I recognize the objective but do not understand its meaning.	I am familiar with the objective but do understand how to relate it to science teaching.	I understand the objective but need more study and experience before I can utilize it fully in science teaching.	I understand the objective and can utilize it fully in science teaching.

		Student Evaluation of Achievement									*Calculated by Teacher*
											Difference
1. A	1	2B	3	4	5	6	7	8E	9		6
B	1	2	3	4B	5	6	7E	8	9		3
C	1	2	3	4B	5	6	7	8E	9		4
D	1	2	3B	4	5	6	7	8E	9		5
E	1	2	3	4B	5	6B	7	8	9		2
2. A	1	2	3B	4	5	6	7	8E	9		5
B	1	2	3	4B	5	6	7	8	9E		5
C	1	2B	3	4	5	6	7	8E	9		6
D	1	2	3	4	5B	6	7	8E	9		3
E	1	2B	3	4	5	6	7E	8	9		5
F	1	2	3	4B	5	6	7	8E	9		4
......											
......											
7. A	1	2B	3	4	5	6E	7	8	9		4
B	1B	2	3	4	5	6	7	8E	9		7

Total Difference in Achievement

Teacher Analysis of the Self-Evaluation Inventory

The above self-evaluation instrument was given near the end of the course. The instructor calculated the difference in achievement before and after taking the course for each objective by tabulating the difference in the "B" and "E" scores. For example, in question 1. A, the student marked "B" near #2 and "E" near #8, indicating a difference of 6 points on the rating scale. In objective 2, however, there is only a difference of 3.

By tabulating or connecting the points of all the "B's" and all of the "E's", the instructor can easily see numerically or graphically how well the student achieved. To determine this student's achievement with other students in the class, the score differences for all objectives could be totaled and compared. A more thorough evaluation can be made, however, if the individual student's achievement is compared with the mean for the class for each objective.

How often should S.E.I.'s be used? As often as the instructor feels necessary. Hofwolt (3) found that multiple use of S.E.I.'s over a quarter does not change the truthfulness of the students. They continue to rate themselves realistically. He further found that the use of the instrument in itself did not seem to change the motivation of students.

How should the instrument be used by the teacher to modify his instruction? An instructor should look over a class' self-evaluational inventories and evaluate each objective. For example, if the majority of students rate little growth in one of the objectives, the teacher then knows that his instruction has not been very effective relative to it. He has, as a result, a base for improving his teaching. The S.E.I. also may be used in counselling students. The instructor can sit down with a student and discuss the rating of each objective with him. The form may also be used in parent-teacher conferences where the parent gets an opportunity to see how his daughter or son rates himself relative to the objectives.

The data from an S.E.I. may be used as a basis for grading. If a student consistently rates himself highly — as having grown in all the objectives — then he should receive a high grade, providing the instructor has no indications that the rating is not realistic. The best judge of how much a student has learned is the student himself, and we should include his judgments in any evaluation we make.

Assessing Student Attitudes About a Unit or a Course

Shown below are several examples of rating scales to assess attitudes. These forms are used to determine how students feel about the subject or a unit. An instructor may also want to evaluate scientific attitudes, for example, does the student suspend judgment until he has enough data? The word attitude therefore may be used in two ways. The evaluation of the student's attitude is discussed in chapter nine.

Covert Mathematics Support Scale

4 – strongly agree with statement
3 – agree with statement
2 – disagree with statement
1 – strongly disagree with statement

(Circle your choice)

4 3 2 1 : 1. I like mathematics more than science.

4 3 2 1 : 2. I like mathematics more than language arts.

4 3 2 1 : 3. I like mathematics more than social studies.

4 3 2 1 : 4. I will never take an elective mathematics course.

4 3 2 1 : 5. One important function of mathematics is to demonstrate the orderliness of the universe.

4 3 2 1 : 6. It is likely that some information in mathematics will be demonstrated to be inadequate or inaccurate in the future.

4 3 2 1 : 7. Mathematical discoveries have done more good than bad for mankind.

4 3 2 1 : 8. Technological advance for the next hundred years will not be as great as the past hundred years.

4 3 2 1 : 9. It is unlikely that a young mathematician will make important discoveries.

4 3 2 1 : 10. Mathematics should remain as a male dominated profession.

4 3 2 1 : 11. If a student is intelligent he should be channeled into a mathematics career because we need mathematicians.

4 3 2 1 : 12. Mathematical discoveries should not be made public if they may become controversial issues.

4 3 2 1 : 13. More Federal support should be made to assist mathematics research.

4 3 2 1 : 14. In the long run, man's lot will not be improved by mathematical knowledge.

4 3 2 1 : 15. Without knowledge in mathematics, man is a miserable creature.

4 3 2 1 : 16. There is so much to be done in mathematics and so little time to do it in.

4 3 2 1 : 17. Mathematics is a "way of thinking."

4 3 2 1 : 18. If something cannot be measured or counted, there is grave doubt that it exists at all.

4 3 2 1 : 19. All miracles have a mathematical or scientific explanation.

4 3 2 1 : 20. The United States should encourage more foreign mathematics students to study in this country (12, 1969).

Mathematics Student Attitudes

In answering this questionnaire please make your judgments on the basis of what these things mean to you in terms of *your* likes and dislikes at the present time. Circle the number which from its assigned meaning most closely represents your feelings toward the concepts.

	Dislike Very Much	*Dislike*	*Undecided*	*Like*	*Like Very Much*
1. Sets	1	2	3	4	5
2. Number sentences	1	2	3	4	5
3. Number line	1	2	3	4	5
4. Decimal notation and place value	1	2	3	4	5
5. Addition	1	2	3	4	5
6. Subtraction	1	2	3	4	5
7. Multiplication	1	2	3	4	5
8. Division	1	2	3	4	5
9. Fractions	1	2	3	4	5
10. Ratios, Percents and Decimals	1	2	3	4	5
11. Real Number System	1	2	3	4	5
12. Informal Geometry	1	2	3	4	5
13. Laboratory Method of teaching Math	1	2	3	4	5
14. Number systems to bases other than 10 (Base 5, Base 2, etc.)	1	2	3	4	5

(11, 1969)

Student Attitudes About Mathematics or Science

Place X in the spaces of your choice

1. good ____:____:____:____:____:____:____ bad
2. reputable ____:____:____:____:____:____:____ disreputable
3. distasteful ____:____:____:____:____:____:____ tasteful
4. pleasurable ____:____:____:____:____:____:____ painful
5. hazy ____:____:____:____:____:____:____ clear
6. optimistic ____:____:____:____:____:____:____ pessimistic
7. important ____:____:____:____:____:____:____ unimportant
8. sweet ____:____:____:____:____:____:____ sour
9. valuable ____:____:____:____:____:____:____ worthless
10. negative ____:____:____:____:____:____:____ positive
11. unpleasant ____:____:____:____:____:____:____ pleasant

12. stale_____:_____:_____:_____:_____:_____:_____fresh
13. nice_____:_____:_____:_____:_____:_____:_____awful
14. complete_____:_____:_____:_____:_____:_____:_____incomplete
15. meaningless_____:_____:_____:_____:_____:_____:_____meaningful
16. tense_____:_____:_____:_____:_____:_____:_____relaxed
17. untimely_____:_____:_____:_____:_____:_____:_____timely
18. wise_____:_____:_____:_____:_____:_____:_____foolish
19. high_____:_____:_____:_____:_____:_____:_____low
20. unsuccessful_____:_____:_____:_____:_____:_____:_____successful

(12, 1969)

SELF-EVALUATION OF AFFECTIVE DOMAIN
ELEMENTARY SCHOOL LEVEL

Student Evaluation Form
for
"There's A Simple Machine In Your Life UNIPAC"

Please complete the form below and drop it in your teacher's box. You don't need to sign your name. *Your* opinion will be appreciated, so don't ask a friend to help you, *please*.

Pleace an X in the appropriate blank.

1. Before you started this UNIPAC, did you like *science*? yes_____ no_____
2. Do you like *science* now? yes_____ no_____
3. Do you like the UNIPAC method of learning, or the required daily assignment method of learning?
   _____UNIPAC _____Required daily assignment
4. What do you like best about this *UNIPAC*?
5. What do you dislike the most about this *UNIPAC*?
6. How could this *UNIPAC* be improved?
7. Rate this UNIPAC with a value from 1 to 5. One (1) means you think this UNIPAC "stinks"; three (3) means you think this UNIPAC is "OK"; and five (5) means you think this UNIPAC is "groovy". Two (2) and four (4) fall between "stinks" and "OK", or "OK" and "groovy."

   Your rating_____ (4)

Evaluation of the Affective Domain, ESCP*

If you were given a choice, which of the following tasks would you perform? Place a #1 in front of your first choice, a #2 in front of your second choice and so on.

_____a. Write a report on Dr. Wesley Powell.

_____b. Read the sports pages of the local newspaper.

_____c. Perform an experiment with the stream table after school.

_____d. Read an article on erosion in *Scientific American.*

_____e. Take part in an after school play.

_____f. Take part in a panel discussion in earth science class on the erosional problems that exist in the local area.

_____g. Make an oral report on soil erosion in the local area.

_____h. Read the comics in the local paper.

_____i. Take part in interscholastic activities.

_____j. Read a book on principles of soil conservation. (15)

**(Earth Science Curriculum Project)*

Evaluation of the Affective Domain for Chemistry

The following are a series of questions allowing you to indicate how you feel about this chapter. PLEASE ANSWER ALL 10 ITEMS. In each case encircle the letter which represents your own reaction to the statement as follows:

SA – if you strongly agree with the statement
A – if you agree but not strongly
N – if you are neutral or undecided
D – if you disagree but not strongly
SD – if you strongly disagree

There are no incorrect answers, and YOUR RESPONSE WILL NOT AFFECT YOUR GRADE.

1. Since becoming familiar with the concept of redox, you have observed any oxidation-reduction reactions outside of the laboratory? SA A N D SD

2. You have made attempts to better understand in more detail chemical changes in your environment since studying this chapter. SA A N D SD

3. You feel you now understand the concepts of chemical and electrical energy. SA A N D SD
4. It is not important to understand how a battery operates just as long as you know how to use it. SA A N D SD
5. Compared to the major concepts in chemistry, redox is a minor idea. SA A N D SD
6. The concept of redox is helpful in understanding atomic structure. SA A N D SD
7. The mathematical formulas relating electrical and chemical energy are difficult to understand. SA A N D SD
8. You understand what is meant by the scientific method. SA A N D SD
9. The work in this chapter increased your understanding of the scientific method. SA A N D SD
10. The attitudes of the scientific method can be applied to problems outside of science. SA A N D SD

(10)

Summary

Instructors have spent considerable time evaluating students, but seldom have asked them how they thought they achieved. Self-evaluational inventories are designed for this purpose. They are instruments for students to use to assess their feelings about themselves (self-esteem), a course, a unit, or attainment of cognitive and affective objectives. The use of S.E.I.s is particularly helpful in evaluating the affective domain. Several research studies have revealed that students do judge themselves truthfully on S.E.I.s, suggesting that wider use of self-evaluational inventories is warranted.

Bibliography

1. Duel, J.H. "A Study of Validity and Reliability of Student Evaluation of Training." Unpublished doctoral dissertation, Washington University, 1956.

2. Fletcher, Jack E. "Teacher-Centered vs. Student-Centered Methods of Large Group Instruction in Elementary Science Methods Classes."

Unpublished doctoral dissertation, University of Northern Colorado, 1969.

3. Hofwolt, C.A. "An Exploratory Study of the Effect of Self-Evaluational Inventories on Student Achievement in High School Science Courses." Unpublished doctoral dissertation, University of Northern Colorado, 1971.

4. Harszy, Arthur J. Ruby S. Thomas Elementary School, Clark County School District, Las Vegas, Nevada. Unpublished paper, 1967.

5. John, M.J. "An Evaluation of End Self-Ratings on Teaching Competencies by Missouri Teachers of Educable Mentally Retarded Children." Unpublished doctoral dissertation, Indiana University, 1965. Cited in *Dissertation Abstracts* XXII, Ann Arbor: University of Michigan, 1967.

6. Johnson, D. and Rising, G. *Guidelines for Teaching Mathematics*. Belmont, California: Wadsworth Publishing Company, 1967.

7. Lowther, M.A. "A Comparison of the Educational Motivation, Self-Evaluation, and Classroom Conduct of High and Low Achieving Eighth Grade Students." Unpublished doctoral dissertation, University of Michigan, 1961. Cited in *Dissertation Abstracts* XXII, Ann Arbor: University of Michigan, 1961.

8. McCormack, Alan J. "The Effect of Selected Teaching Methods on Creative Thinking, Self-Evaluation, and Achievement of Education Methods Course." Unpublished doctoral dissertation, University of Northern Colorado, 1969.

9. Mowers, G.E. "Self-Judgments and Objective Measures as Related to First Semester Academic Achievement of Non-Selected College Students." Unpublished doctoral dissertation, Pennsylvania State University, 1960. Cited in *Dissertation Abstracts* XXX, Ann Arbor: University of Michigan, 1961.

10. Olsen, Don. Adapted from unpublished questionnaire, University of Northern Colorado, 1968.

11. Reince, O.M. "Perception and Achievement in Education Classes for Prospective Elementary School Teachers." Unpublished doctoral dissertation, University of Wisconsin, 1964. Cited in *Dissertation Abstracts* XXV, Ann Arbor: University of Michigan, 1961.

12. Sharpe, Glyn. Unpublished questionnaire, Jefferson County, Colorado, Schools, 1969.

13. Spicola, R.F. "An Investigation into Seven Correlates of Reading Achievement Including the Self-Concept." Unpublished doctoral dissertation,

The Florida State University, 1960. Cited in *Dissertation Abstracts* XXI, Ann Arbor: University of Michigan, 1961.

14. Tillery, B.W. Adapted from "Improvement of Science Education Methods Courses Through Student Self-Evaluation." Unpublished doctoral dissertation, Colorado State College, 1967.

16. Tovrea, Vern. Unpublished questionnaire, University of Northern Colorado, 1968.

Chapter Nine

TEACHER EVALUATION OF THE STUDENT

Evaluation Based on Behavioral Objectives

Valid and reliable evaluation of student achievement, teacher performance, textbooks, tests, and curriculum are a necessity. This chapter deals with evaluating the student, and chapter ten deals with the other areas of evaluation.

Many national projects, such as the Instructional Objectives Exchange at U.C.L.A. and the National Assessment of Educational Progress at Ann Arbor, Michigan, are involved in developing more valid evaluation measures. The objectives of the National Longitudinal Study of Mathematical Ability (N.L.S.M.A.), funded by the National Science Foundation, deal primarily with the relationship of many variables (such as attitude, textbook, teacher background, etc.) to achievement measures in mathematics. "The extent to which these objectives will be met depends fundamentally upon the method and the competency with which mathematics achievement is measured. Thus a major problem . . . has been to decide *what* to measure, *when* to measure it, and *how* it should be measured (6, p. 489)."

Behavioral objectives provide a basis for deciding "*what* to measure" and "*how* it should be measured." Behavioral objectives focus attention on how the student is to perform after instruction. Tests then measure the amount of change undergone by the student. Evaluation is based on how closely the student's performance resembles the performance described in the objective. A well-written objective should be specific enough to suggest a means of measuring achievement. For example, consider the objective:

> Given a right triangle, the student can *construct* an isosceles triangle having one of its congruent sides and an adjacent angle congruent respectively to the hypotenuse and an adjacent angle of the given right triangle. The student can exhibit this behavior with at least three right triangles.

One method of measuring the achievement of this objective might be to draw the right triangle on a piece of paper and give the student a piece of tracing paper with the instructions:

> Using the given triangle, construct an isosceles triangle on the tracing paper. When you finish, you must be able to fit the right triangle over the isosceles triangle so that the hypotenuse fits one of the equal sides and one of the angles of the right triangle matches one of the angles of the isosceles triangle.

Another method might be to give the student a straightedge, a compass, and some paper with the instructions:

> Use the compass and straightedge to construct a right triangle. (Make its area approximately ¼ the area of the paper.) Using this figure, construct an isosceles triangle having the hypotenuse as one of its equal sides. You may extend any side of the right triangle but you may draw only one new line to complete the isosceles triangle.

Either situation could be used to test student achievement. The possibility of inappropriate or invalid achievement measures is lessened.

How achievement is measured is related to the mode of presentation of material. An idea can be expressed in verbal, symbolic or iconic (pictorial) mode. For example, the verbal statement "volume is inversely proportional to pressure" may be expressed by the symbolic formula $V = \frac{k}{p}$—or by the graph below:

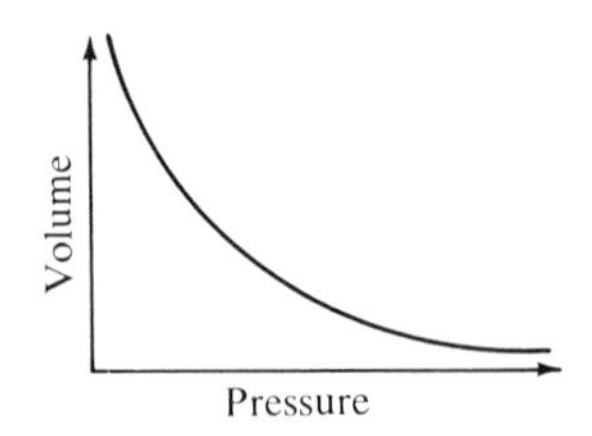

A preference for a particular mode of presentation might affect a student's achievement.

An attempt to obtain information on cognitive preferences has been made by a group of researchers at Sanford through the use of a Cognitive Preference Test (8). The purpose of the test is to determine preference for the symbolic, iconic, or verbal mode of presentation with selected mathematical content.

SELECTED ITEMS FROM A COGNITIVE PREFERENCE TEST

Directions

Mathematical ideas can be expressed in different ways. In this test you are to choose among three *correct* ways of expressing the same mathematical idea.

Each test item consists of three choices dealing with the same topic in mathematics. Read all three of the choices carefully to find the mathematical idea that they have in common.

If you were learning about this mathematical topic, in which of the three ways would you prefer your teacher to explain the idea to you? Mark the corresponding space on the answer sheet.

You may find that more than one choice for each item appeals to you. However, select only *one* choice for each item. Be sure to answer every question, even though the decision may be difficult to make.

A sample item . . . . [follows]

Sample Item

150

a) A number added to three equals eight.

b) $C + 3 = 8$

c)

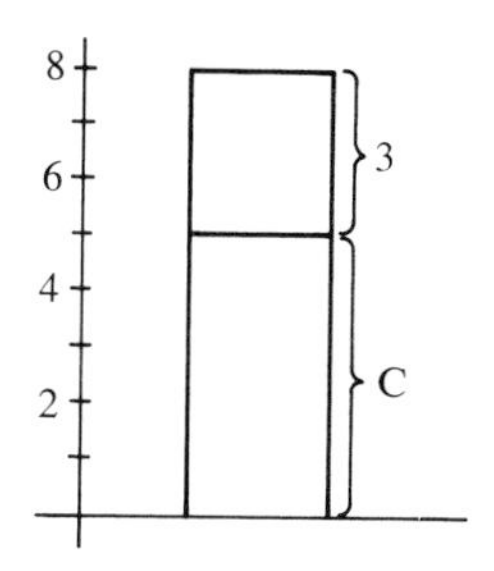

All three choices are correct ways of expressing the same mathematical idea. In which of these three ways would you prefer your teacher to explain the idea? Mark the corresponding space on the answer sheet.

NOTE: For this sample item, blacken the space corresponding to your choice opposite number 150 on the answer sheet.

1

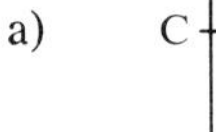

a)

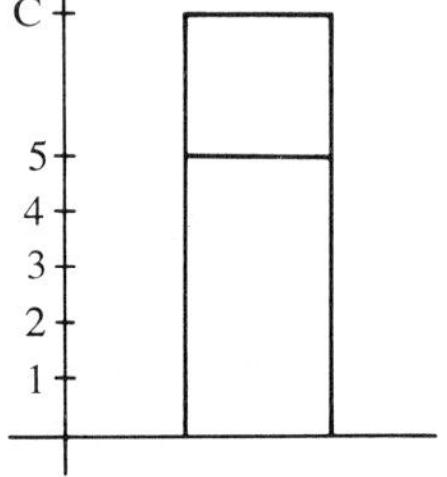

b) Five added to one number equals another number.

c) $5 + T = C$

1. a b c

2

a)

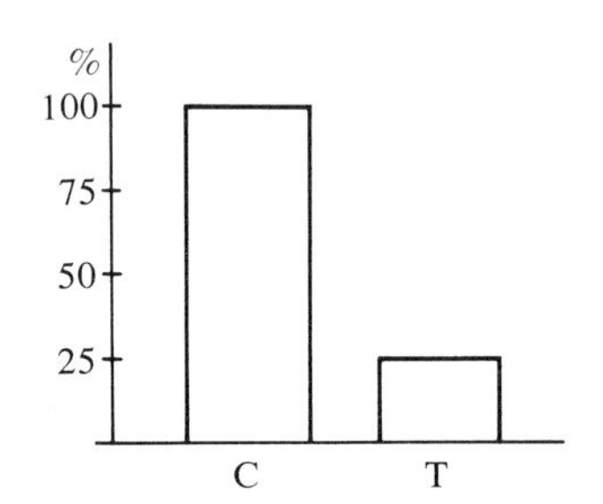

b) $\frac{25}{100} \times C = T$

c) Twenty-five per cent of one number equals another number.

2. a b c

• • • • • •

5.

a)

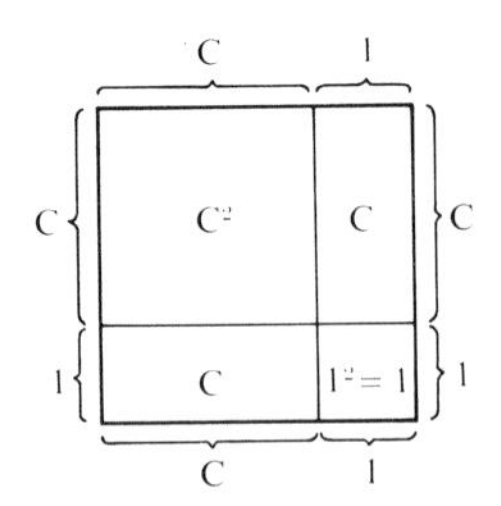

6.

a) The volume of a cube is equal to the length of edge multiplied by itself three times.

b) Volume $= C^3$

where the length of an edge of the cube $= C$

b) When a number plus one is multiplied by itself, it equals that number multiplied by itself plus two times that number plus one.

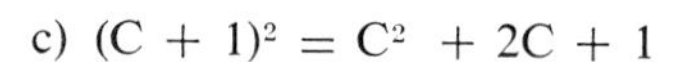

c) $(C + 1)^2 = C^2 + 2C + 1$

5. a b c

c)

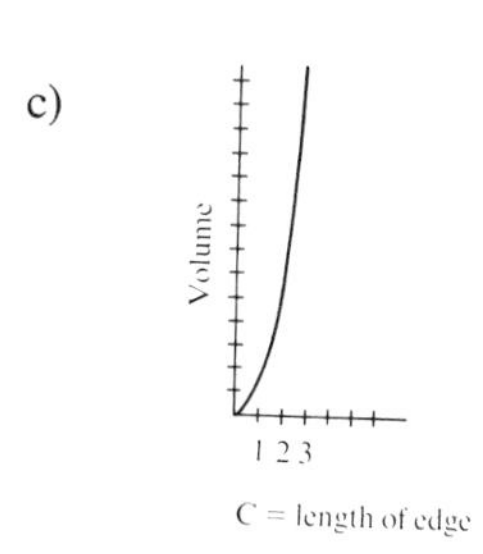

6. a b c

• • • • • •

13.

a) For any right triangle, the square of the length of the hypotenuse is equal to the sum of the squares of the lengths of the other two sides.

b) $\overline{AC}^2 = \overline{BC}^2 + \overline{AB}^2$

where B is the right angle of ΔABC

c)

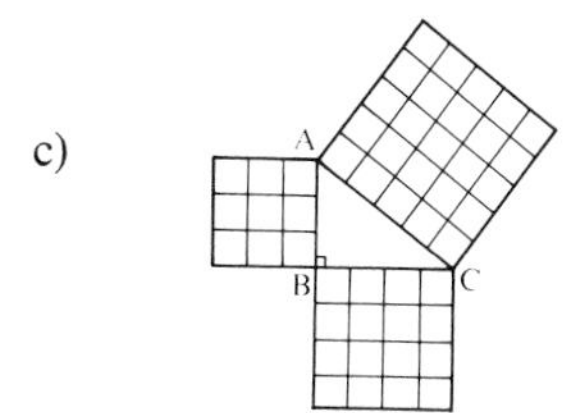

13. a b c

14.

a)

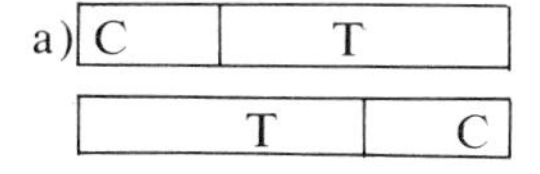

b) For any two numbers, the sum of the first number added to the second number equals the sum of the second number added to the first number.

c) $C + T = T + C$

14. a b c

• • • • • •

21.

a) Perimeter = 2(C + T)

where C = length of rectangle
T = width of rectangle

b)

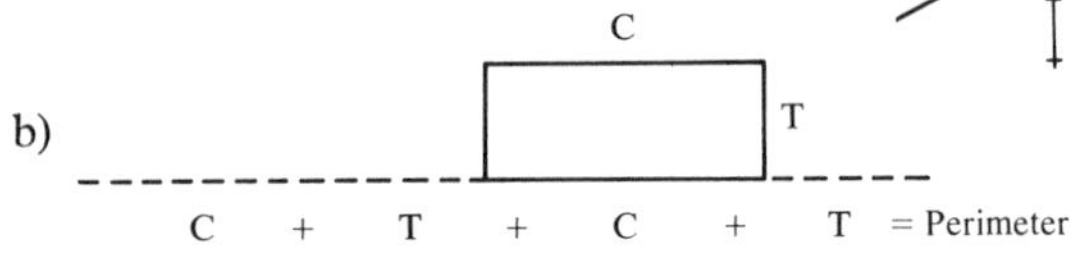

c) The perimeter of a rectangle is twice the sum of the width of the rectangle and the length of the rectangle.

22.

a)

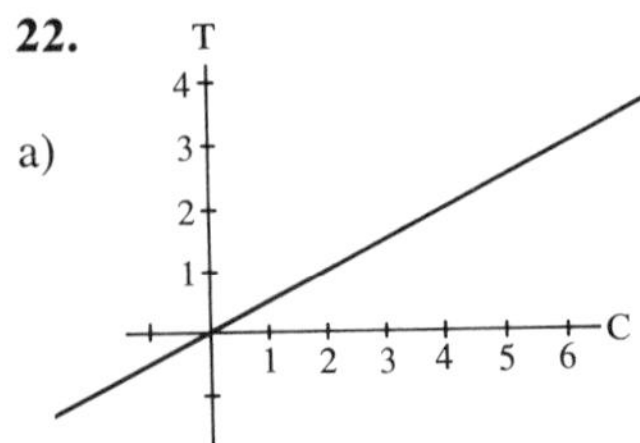

b) T = ½C

c) One number is equal to one-half of another number.

21. a b c

22. a b c

Project CAM (Comprehensive Achievement Monitoring) now at the University of Massachusetts may be able to answer the question of "*when* to measure achievement." The evaluation procedure set up by CAM involves:

1. Instructional objectives and their performance criteria. Each participating teacher must develop specific instructional objectives, enumerating measurable aspects of student behavior and levels of proficiency (performance criteria) desired at the end of the course.
2. Performance tests. In addition to instructional objectives, each teacher will prepare several parallel forms of an instrument to evaluate students' achievement toward the final objectives. In contrast to traditional achievement measures which test one unit of material at a time, Project CAM will monitor the achievement at frequent intervals by evaluating each student's progress against the final criteria set for the course.
3. Initial student achievement. Prior to instruction, one form of the final examination will be given to all students as a diagnostic device. In this way, students will be exposed to major issues and principles and to the way these concepts will be monitored. Teachers will obtain information as to how each student stands on the entire range of performance criteria being considered. Unsuspected competencies and weaknesses will be revealed while there is still time to modify instruction and objectives.

4. Random achievement sampling. In subsequent weeks parallel forms of the final criteria measure will be administered. The sampling procedures are designed to provide:
   a. Valid and reliable sampling of all students performance at the time of each administration.
   b. Adequate monitoring of individual students (1, p. 2–3).

CAM is also attempting to investigate the relation between the student's knowledge of the objectives and his achievement. The behavioral objectives and performance tests written by participating teachers will make CAM an important reservoir of objectives and achievement measures.

The Use of Pictorial Riddles or Non-verbal Problems in Evaluation

Tests of problem-solving ability often measure reading skill and not comprehension. Diagrams, graphs, and pictures, because they are an iconic form of representation, can help students understand problem situations better than the usual written or symbolic forms of communication. Finklestein and Hammil conducted a study to determine how much traditional measures of science achievement evaluated reading ability, rather than science comprehension (6, pp. 34–37). They used the Reading Inventory of Science Knowledge and the Pictorial-Aural Inventory of Science Knowledge. The Science Knowledge test is a regular paper-pencil examination. The Pictorial test is administered by projecting a series of slides on a screen and asking questions about them through the use of a tape recorder. All the student has to do is identify the right diagram. These researchers found that the Science Knowledge test showed expectedly strong dependence on reading. However, when the students were given the Pictorial-Aural Inventory of Science, a parallel form of the knowledge test, there were no significant differences in mean scores between students identified as good readers and those identified as poor readers. The implication of this research is that science and mathematics teachers should guard against the over-emphasizing of tests which are dependent upon the student's reading ability. Any experienced teacher has had students who clearly understood subject matter principles as indicated in discussions, but who flunked questions about these principles on paper-pencil tests. Certainly the use of more pictorial evaluative techniques is warranted.

Some examples of pictorial riddles or non-verbal problems and related objectives are given below.

Objective: When presented with a set of regular geometric figures, the student should *make generalizations* about these figures.

Item: Fill out the following chart and then list all the generalizations or patterns you find. Be sure to test your generalization before you list it.

Figure	Number of Sides	Number of Diagonals	Number of Regions	Sum of Interior Angles
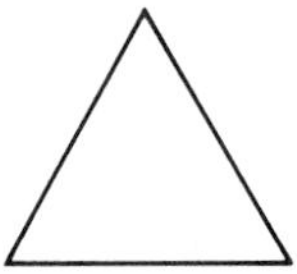				
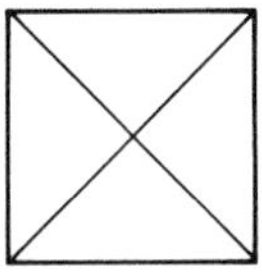				
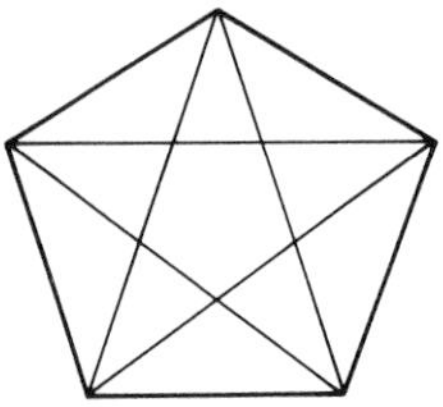				
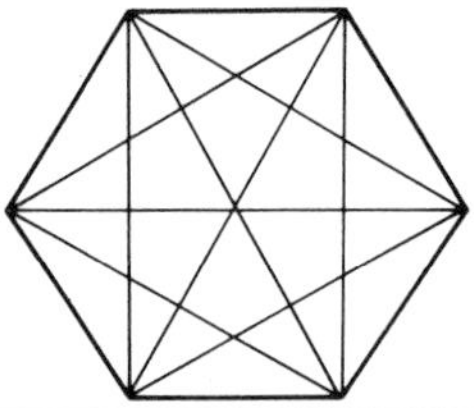				

Objective: When presented with the graphical representation of a set of data, the student should *make* at least three correct *inferences* about the data.

Item: Write three true statements about the situations described by the graphs on page 111.

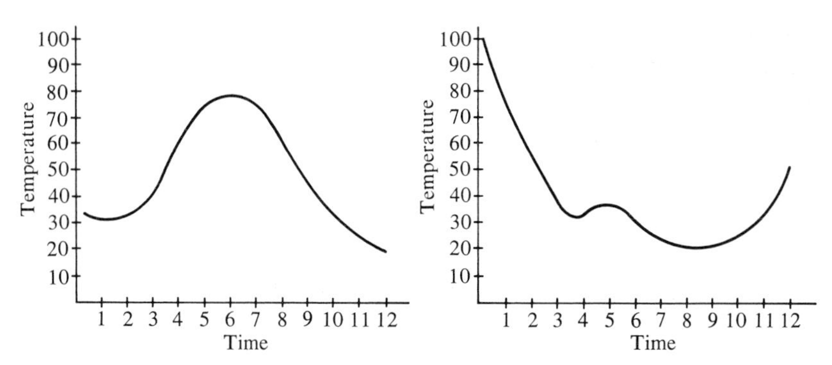

Objective: When presented with appropriate shapes, the student should *construct* a geometric model of a given algebraic statement.

Item: Using only the following shapes, construct a geometric picture of the algebraic statement $(X + 4)^2 = X^2 + 8x + 16$. You may use as many copies of these shapes as you need.

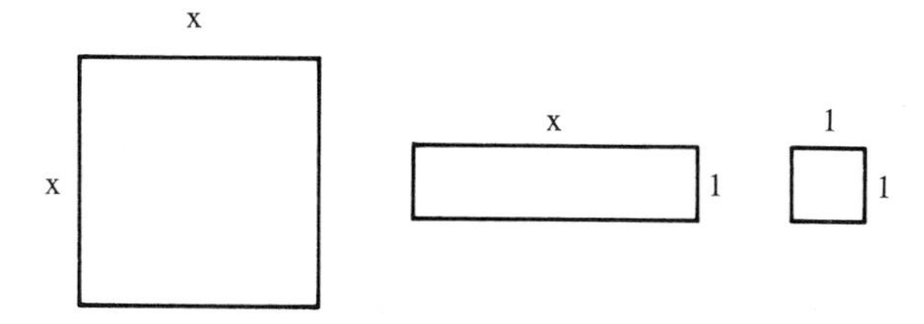

Objective: The student should *construct* an algebraic analog of a given geometric statement.

Item: Based on the information below, predict the number of dots in the tenth picture in this series.

No. of Dots	Difference
1	
	2
3	
	3
6	
	4
10	
	5
15	

Objective: When presented with a picture or diagram containing mathematical data, the student should *construct* problems relating to the picture or diagram which he can solve with the given data.

Item: Make up a problem suggested by the illustration below. Write out both the problem and its solution (13, p. 8).

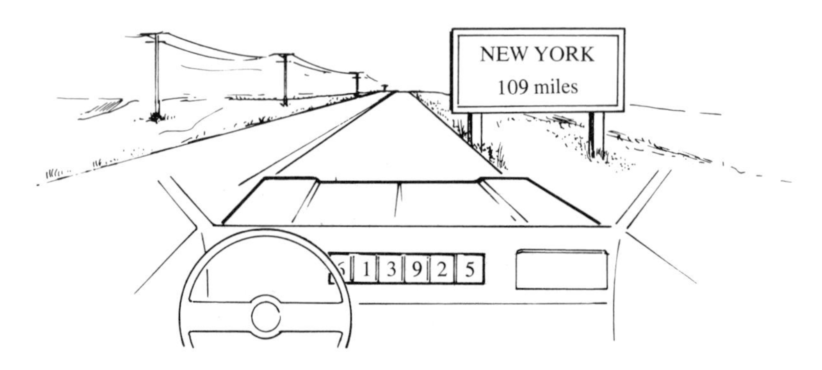

Objective: The student should *apply* the principle that plants are dependent upon animals and animals are dependent upon plants and both of these are dependent upon the environment.

Item:

1. The fish in which tank would survive the longest? Why?
2. In what tank would the plants survive the longest? Why?
3. If you were going to alter tank 1, what would you do and why?

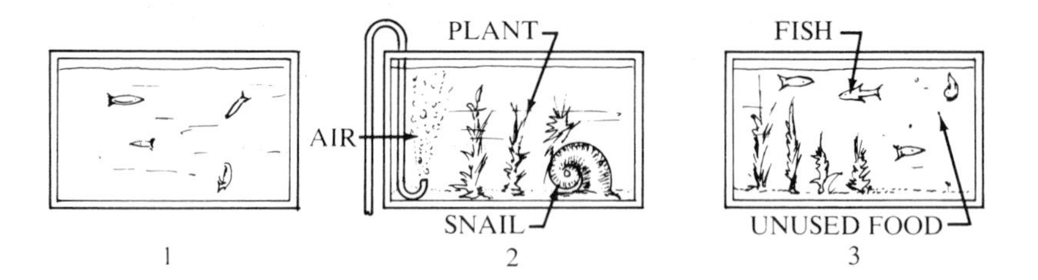

Objectives: The student should:

1. *State* when air is heated in an open container, it expands.
2. *State* when air is cooled in an enclosed container it contracts.
3. *Apply* the principle of how the difference in air pressure may cause an object to move.

Item: Explain what will happen in the situation at the top of page 113.

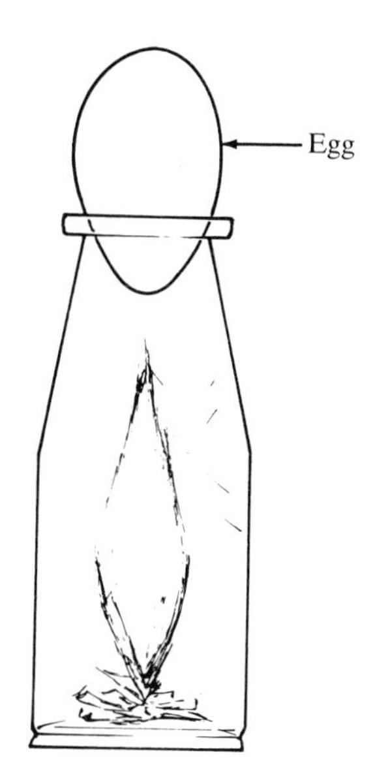

Objective: The student should *explain* why the retrograde motion of Mars occurs.

Item: The diagram below (4) shows the movement of the planet Mars. How would you explain why it appears to move the way it does?

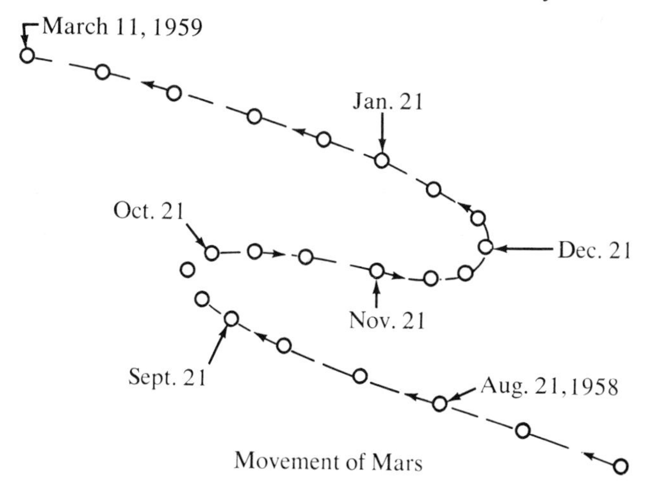

Movement of Mars

Objective: The student should *explain* why and how tides occur.

Item: Explain why the Earth in the diagram on page 114 appears the way it does.

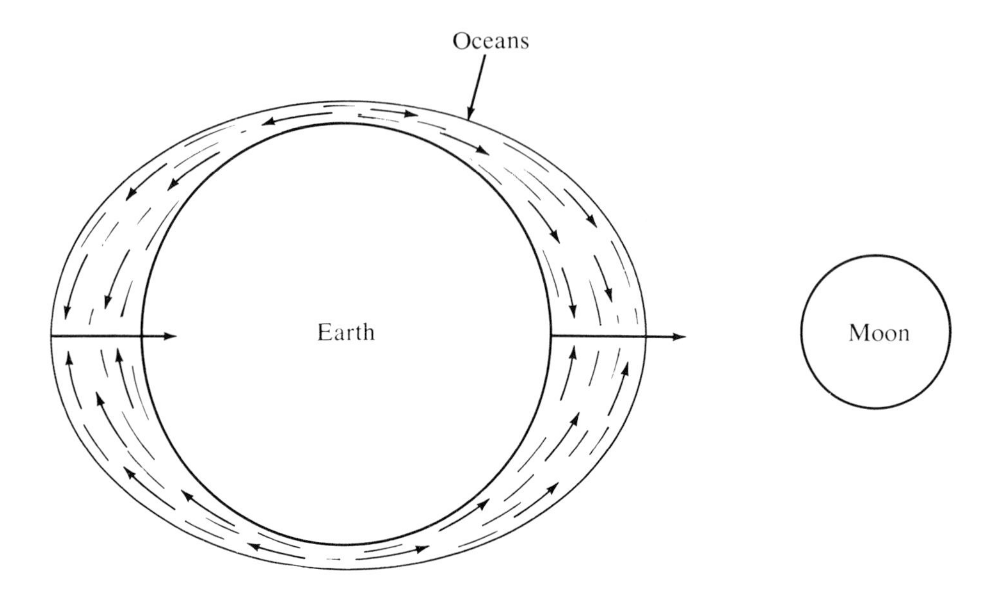

The value of pictorial riddles for evaluating students with reading difficulties is apparent. Of even greater value is their use in evaluating the ability of educable mentally retarded to infer. The following objectives and test items can be used for this purpose.

Patterns

Test Directions:

Examiner reads the questions and puts the possible answers on the board. Show that one answer should be marked with an X. Scoring: 1 point for each correct answer (12). Total: 2 points.

Objective:
When given a series of numbers, the student should *infer* other numbers in the series.

Items:

(a) Here is a pattern: 5 6 7 5 6 7 5 6 7 ?
The next number should be:

(a) 5 (c) 7
(b) 6 (d) 8

(b) Here is a pattern: a b b c d d e f ?
The next letter should be:

(a) e (c) g
(b) f (d) h

(13, p. 73)

STRETCHING

Objective: When given a scale reading, the student should *infer* the number of weights on a rubber band.

Item: Students are shown how a rubber band will stretch if weights are placed on it. They are to note how far one and two weights stretch the rubber band. Then they are asked to place an X next to the number of weights on the fully stretched rubber band shown on the test sheet. Scoring — 1 point for correct answer (d).

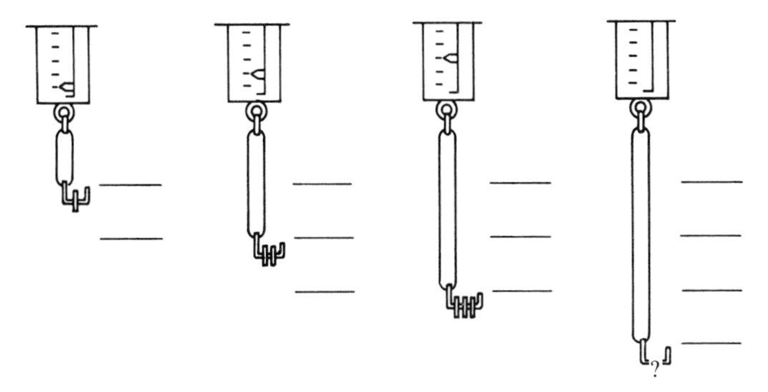

How many weights must be on this?

(a) 1 (c) 3
(b) 2 (d) 4

RISING WATER

Objective: The student should *infer* the correct number of metallic balls in a glass of water by comparing the height of the water in a given vessel with vessels containing a given number of balls.

Item:

a. Students are asked what would happen to the water level in a glass of water if something is dropped inside it. They should note how the water level rises when one and four balls are in the glass, shown in the test sheet. They are asked to write the number of balls they think must be inside the glass in the middle to make the water where it is.

b. Students are told that sailboats placed in a pan of water also make the level rise as shown. They are to write how many boats would bring the level to the mark shown.

Scoring: 1 point for designating 3 balls
1 point for designating 4 boats Total: 2 points.

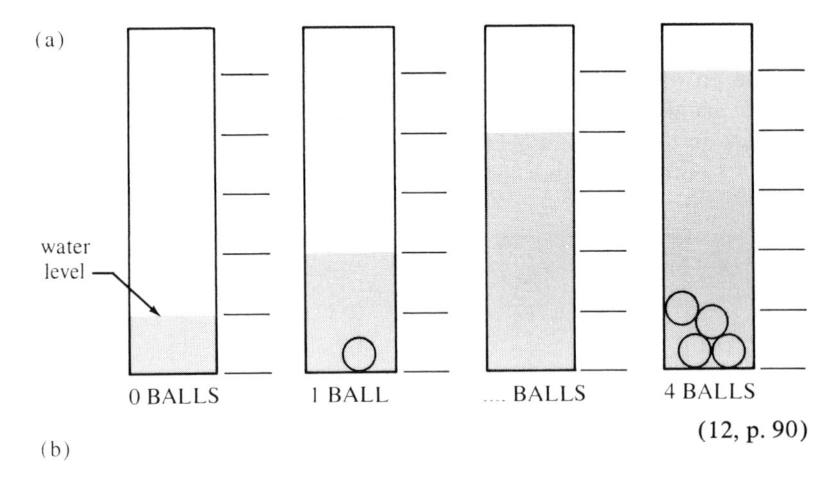

(12, p. 90)

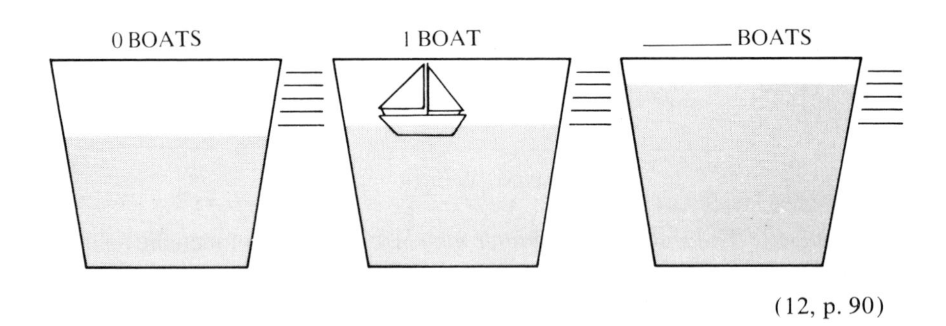

(12, p. 90)

A special form of the pictorial riddle, the Thematic Apperception Test, may be used to determine the student's perception and his desire to apply values. Psychologists use this type of instrument to gain insights into what individuals perceive and believe.

For example, the instructor shows the student a diagram or picture of people working in a science laboratory. The diagram might illustrate several good and bad laboratory practices. The student is asked to respond to the picture in any way he likes. If he points out the bad laboratory practices, explains why they are bad, and suggests how they might be modified, you have an indication that he wants to apply and use what he has learned about good laboratory procedure. However, if a student ignores the examples of poor laboratory practices, you are less certain.

Interview and Observational Techniques

Many teachers avoid interviews and observational techniques because they are time consuming. The data they provide is thought to be less reliable than the "hard" data of traditional evaluation methods. Each of these methods, however, has certain advantages over the paper and pencil method. Some of the advantages of observation are:

1. Observations can be made of the performance of the pupil in a natural, practical situation.
2. Observations permit the pupil to respond without restrictions or tensions that frequently are concomitant with testing.
3. Observations permit the evaluation of certain outcomes that cannot be obtained in any other way.
4. Observations can be continuous and part of the instructional activities.
5. Observations make possible immediate guidance and remedial teaching before undesirable or incorrect habits become established (9, p. 342).

Observation provides a better means of evaluating behaviors in laboratory type situations than do paper and pencil tests. For example, the manner in which a student gathers and assembles equipment, the precision of measurement, and rechecking data for accuracy are important aspects of student performance which lend themselves to observation. Both interviews and observation can be used to obtain data on pre-literate children and on children with reading deficiencies. In addition, interviews permit an in-depth evaluation not possible with other methods.

A checklist is a method of recording data gathered by interview and observational techniques. The one below (4) may be used to record achievement of affective behavior for a science electricity unit. The teacher observes the student in the laboratory and interviews him during his work on the unit. The teacher then records a mark as each of the objectives is achieved.

OBSERVATIONAL ANALYSIS CHART FOR GEOLOGY

Behavioral Response	Students Jack	 Mary	 Robert	 William
1. Shows interest in geological history by asking questions about the history of rocks which are found in his community.				
2. Volunteers to collect rocks from the community to be used in class discussion.				
3. Asks where he might be able to see some of the geological features that have been discussed.				
4. Asks to show to the class his family's slides which were taken during the summer vacation.				
5. Asks to bring some of his rock collections so that the teacher and other students may help to interpret the history of his rocks.				
6. Volunteers to go on a half day field trip for this coming Saturday.				
7. Asks to use some of the teacher's reference books on rocks and fossils to help him interpret the history of some rocks.				
8. Asks the teacher for some good reference books on geological history.				
9. Asks the teacher for a good reference which explains the geological history of the community.				

The following checklist (2) was used to obtain data on fourth-, fifth-, and sixth-grade students and to report achievement. The device consists of two lists. List A contains the course objectives for the semester's work, i.e., a unit on electricity. Evaluation of progress is made by observing the student's success on each of the objectives. As the student performs each task, the appropriate blank is checked. The behaviors vary in difficulty; perhaps none of the students will accomplish the harder ones. List B deals primarily with student attitudes towards the course. No grades are given, nor should any be inferred from the total number of check marks.

SCIENCE LABORATORY
STUDENT PROGRESS REPORT — ELECTRICITY

Section A

1. Can assemble a bulb, battery, and wire in several different ways in order to light the bulb.	
2. Can generalize those factors of circuits in which the bulb lights.	
3. Can construct and demonstrate the use of a simple electrical circuit tester.	
4. Can group a variety of solids and liquids into (a) conductors, (b) non-conductors, of electricity using a simple circuit tester.	
5. Can assemble a circuit from (a) a pictorial representation	
(b) a symbolic representation of that circuit.	
6. Can predict whether a bulb in a given circuit (pictorial or symbolic) will light.	
7. Can convert pictorial circuits into these using electrical symbols.	
8. a. Can make observations on hidden wiring patterns using a circuit tester.	
b. Can use these observations to make inferences about the hidden wiring.	
c. Can draw several possible equally valid inferred wiring patterns to demonstrate an understanding of the fact that the inferred wiring pattern may, or may not, be the actual wiring pattern.	
9. Can make generalizations about the effect of placing certain wiring in an electrical circuit.	
10. Can predict the effect of introducing an additonal wire into a circuit.	
11. Has constructed an original or improved design of simple electrical circuit equipment (e.g. a new bulb holder) for use in this course.	
12. Can demonstrate the meaning of the terms: a. short circuit	
b. complete circuit	
c. open circuit	

SCIENCE LABORATORY
STUDENT PROGRESS REPORT — ELECTRICITY

Section B

1. Bring questions and/or activities to class.	
2. Can work in a group.	
3. Can work independently.	
4. Persists with an area of interest.	
5. Can say "I don't know."	
6. Displays initiative.	
7. Displays skill.	
8. Asks for help when needed.	
9. Refuses help when appropriate.	
10. Asks relevant questions.	
11. Suggests a way of solving a problem.	
12. Challenges ideas, that is, is skeptical.	
13. Contributes a fact.	
14. Contributes an explanation.	
15. Works steadily.	
16. Gets excited about science.	
17.	
18.	
19.	
20.	

(2)

The following rating scales may be used by the teacher from time to time to rate students on their creative involvement, demonstration of science processes, scientific attitudes, and affective progress. These forms could also be modified into self-evaluational inventories for students to rate themselves several times during a course.

OBSERVATIONAL OR SELF-EVALUATIONAL INVENTORY OF STUDENT BEHAVIOR

A. *Creative Processes*	Seldom	Sometimes	Often
1. Formulates problems			
2. Hypothesizes			
3. Designs experiments			
4. Makes inferences			

	Seldom	Sometimes	Often
B. *Other Inquiry Processes*			
1. Records accurate observations			
2. Makes comparisons			
3. Quantifies			
4. Classifies			
5. Collects data			
6. Organizes data			
7. Interprets data			
8. Graphs when needed			
9. Makes Operational Definitions			
10. Identifies assumptions			
C. *Scientific Attitudes*			
1. Demonstrates *objective* attitude by presenting evidence *for* and *against* an idea (Ex. drugs, alcohol, or smoking)			
2. Suspends judgment until he has investigated objectively the subject or states he has insufficient information to make any definite statements			
3. Indicates curiosity about observations by asking questions and making investigations			
4. Indicates in discussion he knows the difference between hypotheses, solutions, facts, inferences by making such statements as: My hypotheses is . . . The solution is . . . The facts are . . . One inference is . . .			
5. Changes opinions when confronted with evidence			
6. States in using statistical data where warranted that although there is a correlation this does not necessarily mean a cause and effect relationship.			
7. States cause and effect relationships			
8. Evaluates his and others procedures and information in experimentation			

The behaviors in the following two check lists should be checked "Frequently," "Seldom" or "Never" for each category by the teacher: Please note this check list would have to be modified slightly for science.

Mathematics or Science Verbal Check List

	Frequently	Seldom	Never
1. *Supports:*			
(a) Proposes desirable actions			
(b) Supports peer's ideas			
(c) Suggests a constructive criticism			
2. *Reports:*			
(a) For more needed information			
(b) For more problems			
(c) For more participation			
3. *Advocates:*			
(a) Why he could not work a problem			
(b) Interprets a peer's solution			
(c) A personal solution to a problem			
(d) A solution to a problem			
(e) A problem by extrapolating			
(f) A special topic or report			
4. *Requires:*			
(a) Problem solution without proof			
(b) Unethical or immoral action			

(11)

Mathematics Non-verbal Check List

	Frequently	Seldom	Never
1. *Participates:*			
(a) In a group mathematics project			
(b) Prepares a special mathematics project			
(c) Writes an article or report on a mathematics topic			

2. *Purchases:*	Frequently	Seldom	Never
(a) Mathematics reading material			
(b) Mathematics materials			
(c) Mathematics-related material			
3. *Listens:*			
(a) To the teacher			
(b) To classmates			
(c) To alternate solutions to problems			
(d) To advise on how to study mathematics			
4. *Borrows:*			
(a) Books and materials from others			
(b) Ideas from teacher and classmates			
5. *Visits:*			
(a) Mathematics centers and displays			
(b) Research and computer centers			
(c) University mathematics departments			
(d) Mathematics fairs			
6. *Selects:*			
(a) Useful mathematics materials from gadgets			
(b) Elective mathematics courses			
(c) A career in mathematics			
7. *Assists:*			
(a) In mathematics activities			
(b) Other students on problems			

8. *Constructs:*	Frequently	Seldom	Never
(a) Geometric models of a mathematics situation			
(b) Mathematical models for demonstration			
(c) Parallel problems or solutions			
9. *Works:* (a) Part-time in a mathematics-related job			
(b) On his own			
(c) With interest			
10. *Reads:* (a) Mathematics journals, articles, etc.			
(b) Mathematics books			
(c) Press releases concerning mathematics			
11. *Receives:* (a) Advice in a constructive way			
(b) Help on problems with appreciation			
(c) Concensus decisions with grace			
12. *Forms Habits:* (a) Completes assignments on time			
(b) Of logical thinking			
(c) Using good problem-solving techniques			
(d) Of cooperation			
(e) Of consideration of others			
13. *Prefers:* (a) Useful instructional materials in laboratory activities			
(b) Using several techniques in solving problems			

14. *Creates:*	Frequently	Seldom	Never
(a) Mathematical models			
(b) Supplementary materials for mathematics			
(c) Alternate solutions to a problem			
15. *Shares:* (a) Materials with classmates			
(b) Ideas in a group			

(11)

Behavior Rating Form

1. Does this student adapt easily to new situations, feel comfortable in new settings, enter easily into new activities?

   _____always _____usually _____sometimes _____seldom _____never

2. Does this student hesitate to express his opinions, as evidenced by extreme caution, failure to contribute, or a subdued manner in speaking situations?

   _____always _____usually _____sometimes _____seldom _____never

3. Does this student become upset by failures or other strong stresses as evidenced by such behaviors as pouting, whining, or withdrawing?

   _____always _____usually _____sometimes _____seldom _____never

4. How often is this student chosen for activities by his classmates? Is his companionship sought for and valued?

   _____always _____usually _____sometimes _____seldom _____never

5. Does this student become alarmed or frightened easily? Does he become very restless or jittery when procedures are changed, exams are scheduled or strange individuals are in the room?

   _____always _____usually _____sometimes _____seldom _____never

6. Does this student seek much support and reassurance from his peers or the teacher, as evidenced by seeking their nearness or frequent inquiries as to whether he is doing well?

   _____always _____usually _____sometimes _____seldom _____never

7. When this student is scolded or criticized, does he become either very aggressive or very sullen and withdrawn?

   _____always _____usually _____sometimes _____seldom _____never

8. Does this student deprecate his school work, grades, activities, and work products? Does he indicate he is not doing as well as expected?

   _____always _____usually _____sometimes _____seldom _____never

9. Does this student show confidence and assurance in his actions toward his teachers and classmates?

   ____always ____usually ____sometimes ____seldom ____never

10. To what extent does this student show a sense of self-esteem, self-respect, and appreciation of his own worthiness?

   ____very strong ____strong ____medium ____mild ____weak

(5, 267–68)

The Mathematics Progress Chart (9, p. 343) below allows the observer to record three degrees of performance — excellent, average, and unsatisfactory.

MATHEMATICS PROGRESS CHART

*Problem, Unit, Activity*____________ *Grade*__________ *Date*______________

Performance Code:

Excellent (†)
Average (0)
Unsatisfactory (-)

	Paul	Mary	Bill
1. Planning activities			
2. Locating information			
3. Using measuring instruments			
4. Organizing information			
5. Recording data			
6. Computing accurately			
7. Cooperating with others			
8. Completing projects promptly			

Jean Piaget, the Swiss psychologist, and his followers have made extensive use of the interview technique to measure the mathematical understandings of children ranging in age from pre-school to adolescence. Knowledge of concepts, such as, one to one correspondence, conservation of number, conservation of volume, and conservation of length have been measured in this way.

D. B. Harrison of the University of Calgary has summarized the procedures of David Elkind in a study of ninety children ranging in age from four to six. The summary provides a model interview situation evaluating discrimination, seriation (ability to arrange in a series), and numeration with three tests involving two sets of nine size-graded sticks being administered to the child.

> In the first test, a discrimination test, the child was presented with one set of nine sticks in disarray on a table. He was given a score of one point for successfully being able to do each of the following things: find the smallest, find the largest, find the smallest after the sticks had been arranged so that the smallest appeared larger than the other sticks, and find the largest after it had been disguised (four points altogether). The second test, a seriation test, involved presenting the child with nine sticks disarrayed and asking the child to make a stairway just like one that the experimenter constructed and then dismantled. If the child was not able to do this with nine, five were removed. If he was successful in making a stairway with four sticks he received one point. If not, he was not tested further. If successful, he was asked to build a stairway with seven sticks (for one more point). If successful again, he was asked to build a stairway with nine sticks (one point). Finally, if he succeeded with nine, five more sticks selected at random from the second set of sticks were brought out and the child was asked if he could put them where they belonged (for one more point). The third test, a numeration test, involved presenting the child with an intact stairway of nine sticks and asking him to count the number of stairs (one point: "How many stairs does the doll have to climb to get on this stair?" If the child was able to answer correctly when each stair in the stairway was pointed to in succession, he was given one more point. A further point was given if the child could answer the preceeding question when the fourth stair was pointed to *and* when the seventh stair was pointed to. Finally, the sticks were mixed and again the fourth and then the seventh stairs were pointed to and the child was again asked how many stairs the doll would have had to climb (for one point). The purpose of giving the numeration test was to determine whether the child could coordinate an ordinal position with a cardinal value (the number of stairs climbed) (8, pp. 14–15).

An excellent demonstration of the interview technique may be observed in two films produced in 1966 by Davidson Films entitled *Piaget Developmental Theory: Conservation* and *Piaget Developmental Theory: Classification.*

Summary

Evaluation is an integral part of education and must be based on the teacher's objectives in order to have validity. Objectives focus attention on how the student is to perform and suggest methods for measuring that performance. The success criteria written into each objective makes it easier to form value judgments about student achievement and hence to assign grades.

Pictorial riddles or non-verbal problems are especially useful in measuring the achievement of slow learners, poor readers, and educable mentally retarded students. Pictorial riddles force average students to consider familiar concepts in a non-verbal form. Interview and observation techniques are best for preliterate students and laboratory-type situations.

The discussion of evaluation in this chapter is certainly not exhaustive. For example, no attempt has been made to discuss the role of teacher intuition and its relevance to the evaluation process. Nor have the relative weights to be assigned to cognitive and affective data been discussed. The emphasis has been on obtaining a greater variety of data and attempting to make the data more relevant to teacher-made objectives.

Bibliography

1. Allan, D., et al. "Comprehensive Achievement Monitoring." Mimeographed. Stanford, California: Project CAM, School of Education, 1968.
2. Allen, Leslie R. "Science Laboratory Student Progress Report: Sections A and B." University Laboratory School, University of Hawaii.
3. Carlson, Roger. Unpublished paper, University of Northern Colorado.
4. Champion, Ruth C. Unpublished paper, University of Northern Colorado.
5. Coppersmith, S. "The Antecedents of Self-Esteem." San Francisco: W.H. Freeman and Company Publishers, 1967, pp. 267–68.
6. Finkelstein, B. and Hammill, D. "A Reading-Free Science Test." *The Elementary School Journal,* October 1969, pp. 34–7.
7. Harrison, D.B. "Piagetian Studies and Mathematics Learning and Instruction." Paper presented at the forty-seventh annual meeting of the National Council of Teachers of Mathematics, Minneapolis, Minnesota, April 1969.
8. Heath, Robert and Snow, Richard. "Cognitive Preference Test." Mimeographed. Palo Alto, California: Stanford University.
9. Johnson, D. and Rising, G. *Guidelines for Teaching Mathematics.* Belmont, California: Wadsworth Publishing Co., Inc., 1967.
10. Romberg, T. and Wilson, J. "The Development of Mathematics Achievement Tests for the National Longitudinal Study of Mathematical

Abilities." *The Mathematics Teacher,* vol. 61, no. 5, May 1968, pp. 489–95.

11. Sharpe, Glyn. Unpublished scales. Jefferson County, Colorado, Schools.

12. Sweeters, W. "Discovery Oriented Instruction in Science Skills for Mentally Retarded Children." Appendix, doctoral disseration, Colorado State College, 1968.

13. Trueblood, C.R. "Promoting Problem-Solving Skills Through Non-Verbal Problems." *The Arithmetic Teacher*, vol. 16, no. 1, January 1969, pp. 7–9.

CHAPTER TEN

SAMPLE COGNITIVE OBJECTIVES AND ACHIEVEMENT MEASURES

Elementary School Mathematics

The following set of objectives and test items were prepared for the Individually Prescribed Instruction Program at the University of Pittsburgh and are intended for five and six year olds.

ADDITION

Prerequisites: A Numeration

Task Categories:

1. Joining sets and indicating how many.
2. Associating compound addition numerals with the joining of two sets (by selecting numerals, selecting and constructing sets).
3. Associating compound addition numerals with taking steps on a number line segment.

Objectives	*Sample Test Items*
1a. *Puts* two sets *together* (pastes pictured sets, folds page, draws ring around all, shades in common background).	1a. "Put the two sets together by drawing a ring." "Put the two sets together by folding on the dotted lines."

1b. Indicates *how many* things are in each of two sets and then how many things there are *altogether*.

1b. "How many bugs are on the line?"
"How many owls are on the line?"
"How many animals are on the line altogether?"

2a. Indicates that the component numeral of an addition numeral are counting numerals.

2a. "X the counting numerals:
1 2 + 2
Answer 1̸2̸ + 2

2b. Presented with two sets with 0-5 things each, selects a compound *addition numeral* in response to being asked *how many* things there are altogether.

2b. "Circle the addition numeral for how many triangles you see."
1 +1 3 + 2 5 + 0

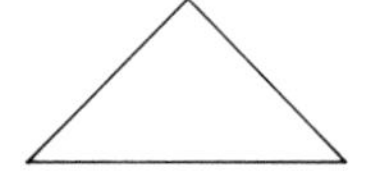
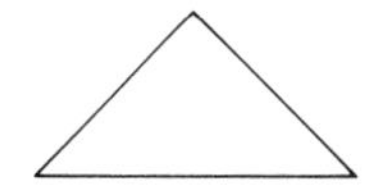

2c. *Makes tally marks* above each counting numeral in a compound addition numeral, numerals 0 + 0 through 5 + 5. Responds to the command, e.g., "Make four plus three tally marks (present numeral)."

2c. "Make this many tally marks."
3 + 5
Answer ''' '''''
3 + 5

2d. Given a compound addition numeral whose sum is 5 or less, makes that many *tally marks* and then selects the counting numeral which represents *how many tally marks* there are altogether.

2d. "Make this many tally marks."
2 + 5 "Circle the numeral which says how many tally marks there are altogether."
0 5 4

2e. Selects the set with as many things as represented by a given compound numeral.

2e. "Put an *X* next to the set with *this* many stars."

2 + 1

2f. Selects the compound addition numeral that is equal to a given counting numeral.

2f. Teacher points to each compound numeral in turn saying, "Make this many tally marks." "Now circle the addition numeral for this many tally marks."

/ /	/ / / /
1 + 1	3 + 1

Answer 4

(5, Table 2)

Geometry in the Elementary School

The inclusion of geometry into the elementary school curriculum is one of the more significant of the current reforms. Topics such as symmetry, congruence, similarity, shape and size, transformations (translations, reflections, and rotations), coordinate systems and graphs, and constructions are included in many new elementary texts. The following objectives and test items sample a few of these topics.

Shape and Size

Objective: When presented with physical objects, such as spheres, rectangular prisms, and cylinders, the student should identify distinguishing characteristics of these objects.

Item: The child is shown a ball, a cereal box, and a tin can and asked to tell everything he can about them using the words, corner, edge, and side.

Objective: When presented with models of common polyhedra, the child is able to verify Euler's formula $V + F = E + 2$ where V is the number of vertices, F is the number of faces, and E is the number of Edges.

Item: Use the set of solids provided to complete this chart. What conclusions can you make from the information below [page 134]?

Solid	No. of Vertices, V	No. of Faces, F	No. of Edges, E	V + F − E
Cube		6		2
Tetrahedron	4			
Octahedron			12	
Dodecahedron				2
Icosahedron		20		
Cereal Box	8			
Shoe Box			12	

Symmetry

Objective: When presented with a figure drawn on paper, the child is able to fold the paper in order to determine a line of symmetry.

Item: For each of the drawings, fold the paper so that the two halves fit on top of one another. You can check to see if they fit by holding the paper up to the light. After you are sure they fit, open up the paper and darken the fold with your pencil.

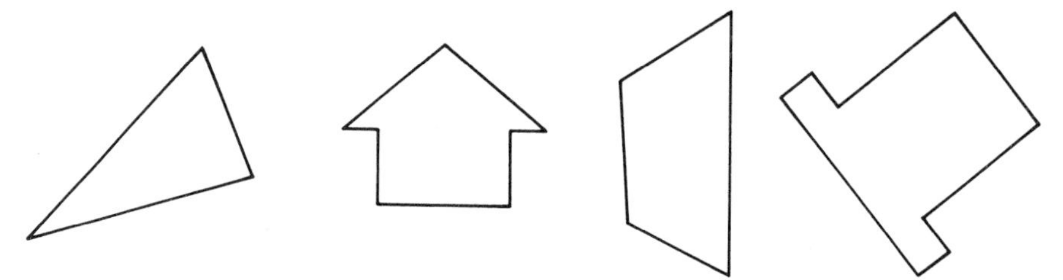

Objective: When presented with a partially incomplete figure and a line of symmetry for the figure, the child can complete the figure.

Item: A line of symmetry and part of a word are shown below. Try to complete the word. You may use tracing paper if you want. Write the complete word of this line ______________________.

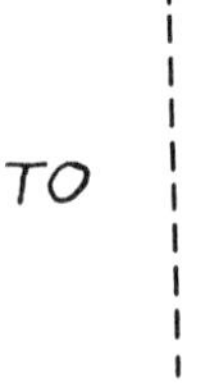

Congruence

Objective: When presented with plane figures and tracing paper, the child can identify congruent figures.

Item: Use your tracing paper to find the two dogs that are exactly alike. When you have found them, circle them.

(6, p. 1)

Objective: When presented with a plane figure, the child can construct a figure congruent to the given one using a distance preserving transformation, i.e., a translation, a refection, or a rotation.

Item: Arturo is an acrobat. First he slid, as shown by the arrow. From his "new" position he flipped about the given flip line. After that, he made a half turn about a turn center right in the middle of his body, followed by this slide: ⟶

Draw Arturo's image after he performs each of these motions (6, p. 84).

Similarity

Objective: When presented with a set of plane figures, the child can identify those which are similar.

Item: Circle the small figures below which could be made to match the large figure if they were enlarged or shrunk with an overhead projector.

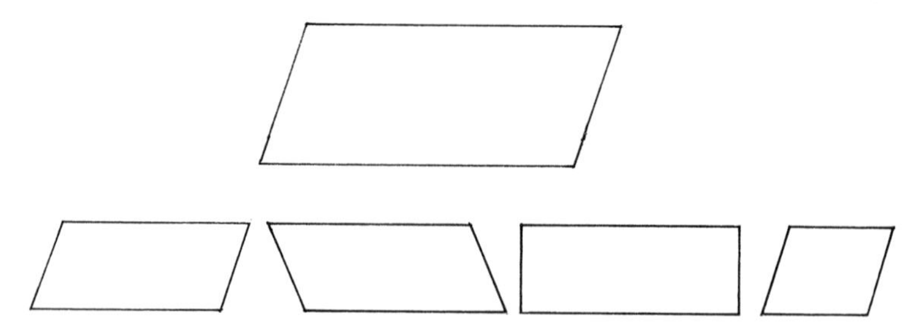

Objective: When presented with a pair of similar figures and the ratio between them, the child can compute the length of indicated segments.

Item: The floor plan on the right is a 3 to 1 enlargement of the one on the left.

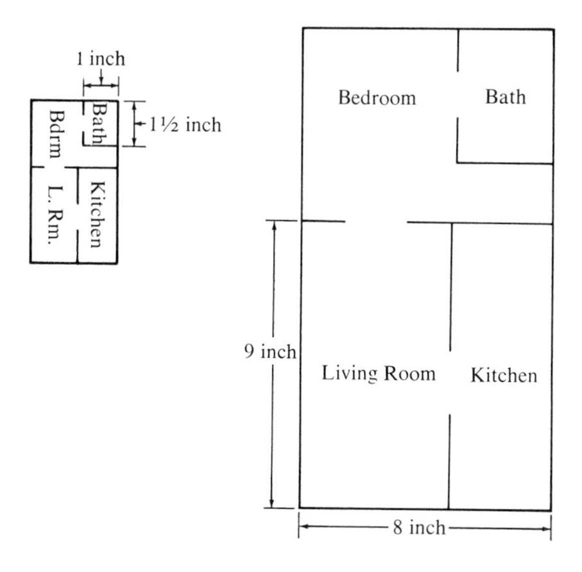

1. In the enlargement, what are the dimensions of the bathroom? ______________.

2. What is the larger dimension of the living room in the original? ______________.

Intermediate School Mathematics

Number, Numeral, Numeration Systems

Objective: The student will *translate* a given verbal statement into a symbolic statement.

Item: A symbolic statement for the sentence, "Three-fourths of some number is one half of fifty-six," is:

(a) (circled) $N - \frac{1}{4}N = \frac{1}{2}(56)$ (c) $2(\frac{3}{4} \div N) = 56$
(b) $\frac{3}{4} \div N = \frac{1}{2}(56)$ (d) $N - \frac{3}{4} \bullet N = \frac{1}{2}(56)$

Objective: The student will *choose* the nearest approximation to a given number.

Item: The closest approximation to the number 2.314×10^{-2} is:

(a) 230 (b) 23 (c) 2/10 (d) .02 (e) −230

Objective: The student will approximate the location of a given number on a given number line.

Item: Darken two consecutive markings on the number line below so that the number 2.37 is between them.

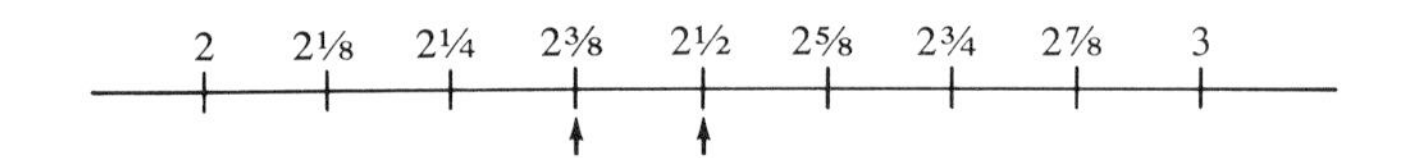

Operations and Their Properties

Objective: The student will *construct* a geometric model of a given problem involving multiplication of fractions.

Item: Use a rectangular model to represent the product 2¼ x ⅓.

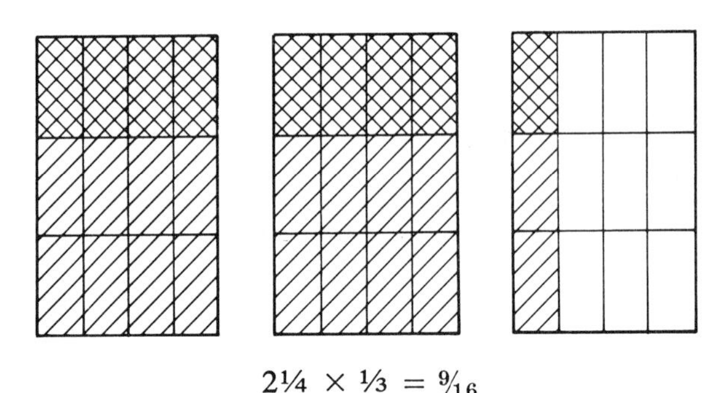

$2\frac{1}{4} \times \frac{1}{3} = \frac{9}{16}$

Objective: The student will *select* the numerical statement equivalent to a given number line representation.

Item: Circle the letter of the numerical statement equivalent to the situation pictured on the number line.

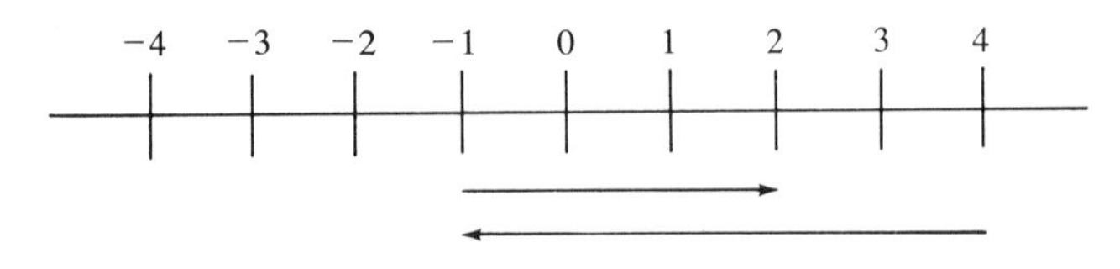

(a) 3 − 5 (b) 2 − (−1) (c) −5 + 3 (d) 4 − 5 + 3 (e) −1 + 4

Functions and Graph

Objective: The student can *name* the domain and the range of a function f when given a graph of f.

Item: The function f is defined by the graph below. Write the set of elements in the domain. Write the set of elements in the range.

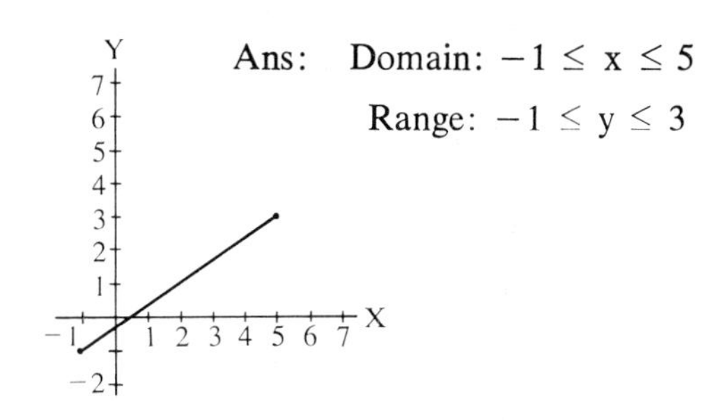

Objective: The student can *extrapolate* the graph of a given function f for a given subset of the domain.

Item: A part of the graph of $f(x) = x^3$ is given below.

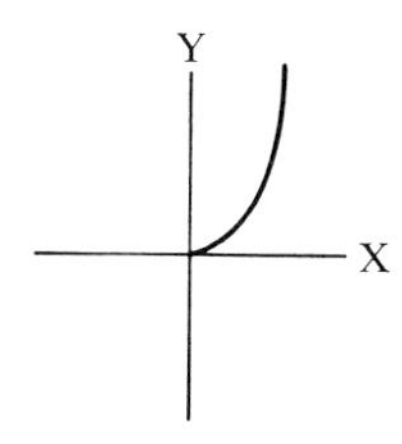

The domain of f is the set of reals.
If x is a negative number then f(x)

(a) is negative and increasing
(b) is negative and decreasing
(c) is positive and increasing
(d) is positive and decreasing
(e) is not defined.

Geometry

Objective: The student can *calculate* the area of a given trapezoid in two ways:

(a) By decomposing it into two triangles
(b) By decomposing it into a rectangle and two triangles

Item: Calculate the area of the trapezoid below by using two different decompositions.

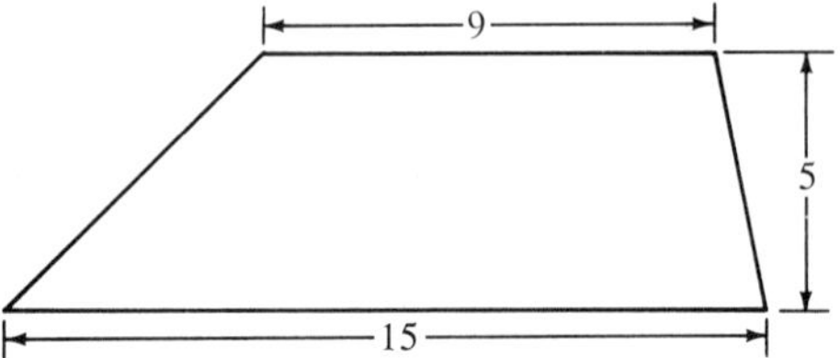

Objective: When presented with plane figures having a line of symmetry, the child can verify the congruence of segments and angles by using the line of symmetry.

Item: In the figure below, the dotted line is a line of symmetry. You may use your mirror or fold the paper to answer the questions.

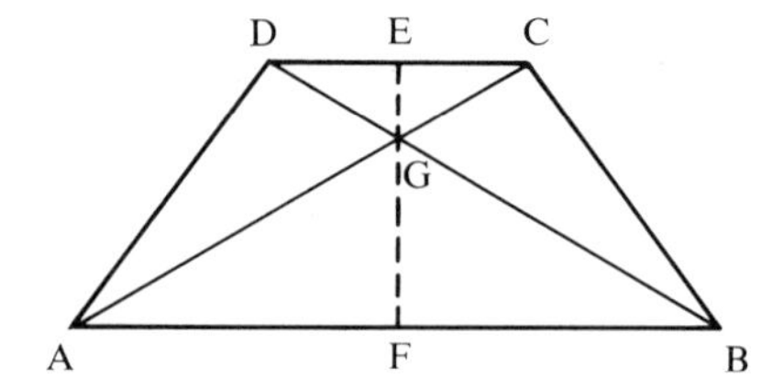

(a) Segment AF is congruent to segment____________________.

(b) Segment DG is congruent to segment____________________.

(c) Angle ADE is congruent to angle____________________.

Measurement

Objective: The student will *compare* measurements using a standard and an arbitrary unit.

Item: A length of one inch is marked off on the line segment below. Next to it, mark off a unit of your own such as the width of your pencil or the length of your thumb. Would the measure of the length of this paper be greater or less with your unit than with the 1″ unit? Check your answer by measuring   Ans: Less

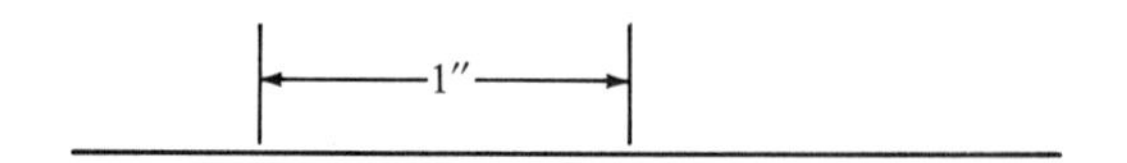

Objective: The student will *calculate* an indirect measurement by using similar triangles.

Item: What is the length of the shorter vertical segment? Ans: 5

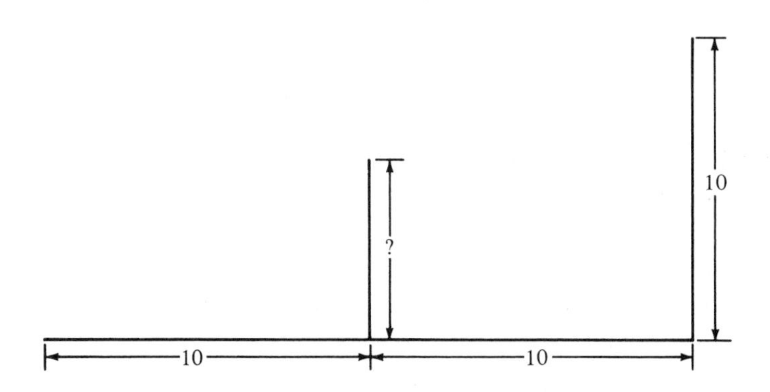

Probability and Statistics

Objective: The student will *predict* the likelihood of an event when given a graph of the data.

Item: The graph below represents the average monthly rainfall on the island of Ponape.

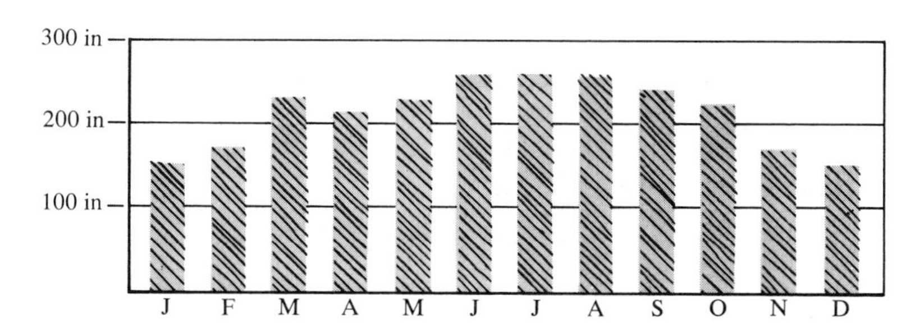

(a) If you wanted to be there when the weather was least wet, in which months would you schedule your visit?

(b) What are the chances of being there during a month with more than 200 inches of rain?

Ans: (a) Dec., Jan., Feb. (b) $7/12$

Objective: The student will *compute* the probability of a given event.

Item: The probability of picking an ace from a standard deck of playing cards is

(a) $1/4$ (b) 4 (c) $1/13$ (d) 1 (e) $1/52$

Secondary School Mathematics

The following are examples of objectives and related test items for various high school mathematics courses.

Trigonometry

The following objectives and test items are taken from a set of objectives written for a two semester trigonometry course.

Objective: To state the value of any of the six trigonometric functions of an angle in standard position, given the coordinates of a point on the terminal side of the angle.

Item: Given $(-5, -12)$ a point on the terminal of $\angle X$; $\therefore \cos X =$

(a) none of the answers below.
(b) $12/5$
(c) $5/12$
(d) $-12/13$
(e) $-5/13$

Objective: To state the value of any of the remaining five trigonometric functions of an angle in standard position, when given the value of one of the trigonometric functions of that angle and the quadrant in which the terminal side of the angle lies.

Item: If $\tan X = -8/15$, $X < 270°$, then $\sin X$ equals:

(a) $-15/8$
(b) $8/17$
(c) $-15/17$
(d) $15/17$
(e) $-8/17$

Objective: To state the value of any of the six trigonometric functions of an angle in standard position when given the measure of the angle, if its reference angle is 0°, 30°, 45°, 60°, or 90°.

Item: $\sec 60° =$

(a) 0.8660

(b) 1.7321
(c) 0.500
(d) 2.000
(e) none of these

Objective: To determine the trigonometric functions of angles whose reference angles are not found in a Table of Trigonometric Functions by linear interpolation, to the number of significant digits determined by the table used.

Item:

	sin	cos	tan	ctn	sec	csc	
39°00′	.6293	.7771	.8098	1.235	1.287	1.589	51°0′
10′	.6316	.7753	.8146	1.228	1.290	1.583	50′
20′	.6338	.7735	.8195	1.220	1.293	1.578	40′
30′	.6361	.7716	.8243	1.213	1.296	1.572	30′
40′	.6383	.7698	.8292	1.206	1.299	1.567	20′
50′	.6406	.7679	.8342	1.199	1.302	1.561	10′
40°00′	.6428	.7660	.8391	1.192	1.305	1.556	50°0′
	cos	sin	ctn	tan	csc	sec	

Using the table above, evaluate sin 309°47′.
(a) 0.6323
(b) 0.7683
(c) 0.7711
(d) −0.6323
(e) −0.7685

Objective: To identify those trigonometric functions which are increasing functions, and those which are decreasing functions, for any specified domain.

Item: If 180° <x <360°, then:
(a) sin x increases in that domain.
(b) tan x decreases in that domain.
(c) cos x increases in that domain.
(d) sin x decreases in that domain.
(e) none of these.

Item: If 270° <x <450°, then:
(a) none of the below is correct.
(b) sec x decreases in that domain.

(c) sin x decreases in that domain.
(d) cos x increases in that domain.
(e) tan x increases in that domain.

Objective: To determine the trigonometric functions of angles greater than 90° whose reference angles can be found in a Table of Trigonometric Functions, and permitted the use of such a table when necessary.

Item: sec 300° =
(a) -2
(b) $-\sqrt{3}$
(c) 2
(d) ½
(e) $-\sqrt{3/2}$

Item: csc 135° =
(a) -1
(b) 1
(c) $-\sqrt{2}$
(d) $\sqrt{2}$
(e) $\sqrt{2/2}$

Item: cos 210° =
(a) $-½$
(b) $-\sqrt{3/2}$
(c) $\sqrt{3/2}$
(d) 1.7321
(e) none of these

Objective: To compute the linear distance between two points, given their rectangular coordinates, using the Pythagorean Theorem.

Item: The distance between the points (12, −9) and (5, 15) is:
(a) none of the below
(b) $5\sqrt{13}$
(c) 625
(d) 15
(e) 25

Objective: To compute the linear distance between two points, given their polar coordinates $(r_1, \Theta 1)$ and (r_2, Θ_2), using the Law of Cosines.

Item: The distance between the points (3, 67°) and (4, 127°) is:
(a) none of the below
(b) 5
(c) 13

(d) $\sqrt{37}$
(e) $\sqrt{13}$

Objective: Given a trigonometric expression containing more than two different trigonometric functions of the same variable angle, to substitute appropriate identities in terms of one of the given functions, and performing the indicated computations, resulting in an expression involving only that one trigonometric function.

Item: $\dfrac{(\sin x)(\cos x)(\tan x)}{\csc x} =$

(a) 1
(b) $\sin^3 x$
(c) $\tan^3 x$
(d) $\cos^3 x$
(e) none of these

Item: $\dfrac{\csc x \sqrt{\csc^2 x - 1}}{\text{ctn}^2 x + 1} =$

(a) none of the below
(b) sec x
(c) ctn x cos x
(d) sin x
(e) cos x

Objective: To verify trigonometric identities in one angle, by transforming one member of the given equation by substituting appropriate trigonometric identities for the functions not mentioned in the other member of the equation, and performing all computations, resulting in identical members in the equation.

Item: Verify this identity:
$\sin \Theta + \cos \Theta \,\text{ctn}\, \Theta = \csc \Theta.$

Item: Verify this identity:

$$\sec^2 x - \sin x \tan x + \tan x \csc x = \frac{1 + \cos^3 x}{\cos^2 x} \qquad (12)$$

Vector Geometry

The following objectives and test items were constructed for a one semester course in vector geometry.

Objective: Given a set of non-collinear points, the student can first name the line segments determined by the set of points and then name the directed line segments determined by these points.

Item:

(a) List all the line segments determined by points A, B, C.

(b) List all directed line segments determined by points A, B, C.

●A

●B

●C

Item:

Given ABCD is a parallelogram. Name:

(a) All the line segments determined by the vertices of the parallelogram

(b) All the *directed* line segments determined by the vertices of the parallelogram

Objective: Given a set of directed line segments in a coordinate plane along with the coordinates of their tail and head points the student can represent vectors by using the notation: $\overrightarrow{PQ} = (e, f)$ where $e = x_Q - x_P$,
$f = y_Q - y_P$

Item:

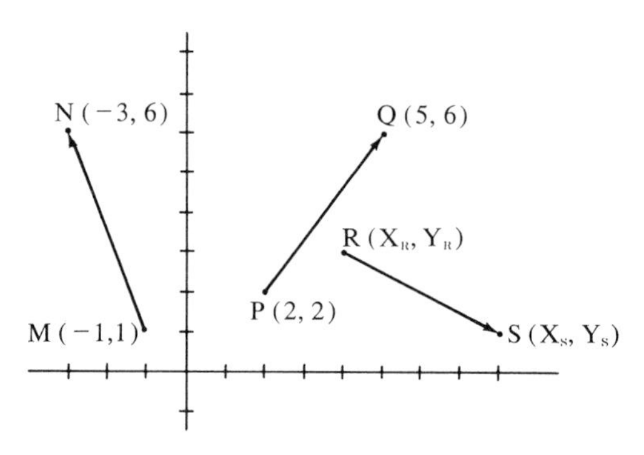

(a) Represent vector PQ in the form $\overrightarrow{PQ} = (e, f)$

(b) Represent vector RS in the form $\overrightarrow{RS} = (e, f)$

(c) Represent vector MN in the form $\overrightarrow{MN} = (e, f)$

Objective: Given two vectors that have the same direction, not necessarily the same sense and not necessarily the same magnitude, the student can correctly identify the scalar in the vector equation that describes one as a scalar multiple of the other. For example

A ⟶ B

C ⟶ D

If $\overrightarrow{AB} = K\,(CD)$
The student can write
$K = 3$

Item: ABDC is a parallelogram, E is the mid point of $\overline{CD}$, and F divides $\overline{EB}$ and $\overline{DA}$ in the ratio 1:2

Find k so that the following statements will be true.

(a) $k(\overrightarrow{AB}) = \overrightarrow{CD}$
(b) $k(\overrightarrow{AD}) = \overrightarrow{AF}$
(c) $k(\overrightarrow{EB}) = \overrightarrow{EF}$
(d) $k(\overrightarrow{CD}) = \overrightarrow{EC}$

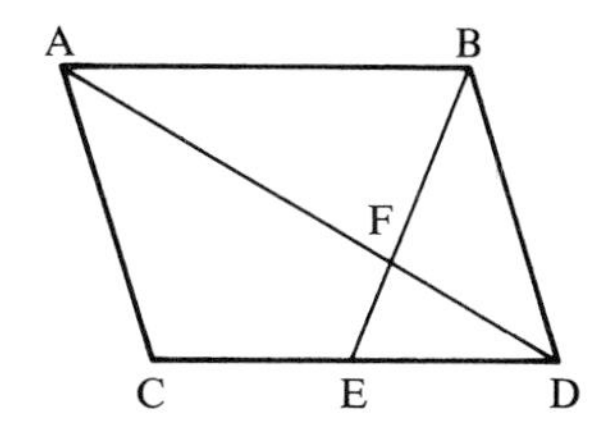

Objective: Given any three of the five variables: coordinates of P, coordinates of Q, (e, f) notation of $\overrightarrow{PQ}$, x component of $\overrightarrow{PQ}$, y component of $\overrightarrow{PQ}$, the student can correctly write the other two.

Item: Fill in the empty spaces:

Coordinate P	Coordinate Q	(e, f) of $\overrightarrow{PQ}$	x component of $\overrightarrow{PQ}$	y component of $\overrightarrow{PQ}$
(1, −3)			−2	4
	(5, 3)		3	−1
(5, 1)	(−2, −1)			

Objective: Given $\vec{a} = (e, f)$, $\vec{b} = (g, h)$ the student can correctly calculate the sum of $\vec{a}$ and $\vec{b}$ as $\vec{a} + \vec{b} = (e + g, f + h)$ and alternately, given the (e, f) notations of $\vec{a}$ and $\vec{a} + \vec{b}$, the student can write the (e, f) notation for $\vec{b}$.

Item: Given $\vec{a} = (-3, 1)$ $\vec{b} = (5, 2)$ $\vec{c} = (4, -3)$

Find:

(a) $\vec{a} + \vec{b}$
(b) $\vec{b} + \vec{a}$
(c) $(\vec{a} + \vec{b}) + \vec{c}$
(d) $\vec{a} + (\vec{b} + \vec{c})$
(e) vector (x, y) if $(x, y) + (-3, 1) = (5, 2)$

Objective: Given a sketch of $\overrightarrow{PQ}$ and $\overrightarrow{QR}$ and their (e, f) notations the student can sketch the vector that represents the difference $\overrightarrow{PQ} - \overrightarrow{QR}$ and express it in the correct (e, f) notation.

Item: Sketch the vector $\overrightarrow{PQ} - \overrightarrow{QR}$ and express it in (e, f) notation

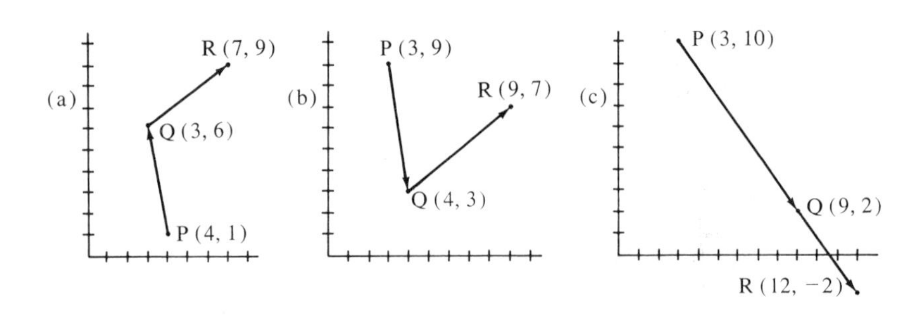

(a) $\overrightarrow{PQ} = (-1, 5)$
$\overrightarrow{QR} = (4, 3)$
$(\overrightarrow{PQ} - \overrightarrow{QR}) = (\_\_\_\_)$

(b) $\overrightarrow{PQ} = (1, -6)$
$\overrightarrow{QR} = (5, 4)$
$(\overrightarrow{PQ} - \overrightarrow{QR}) = (\_\_\_\_)$

(c) $\overrightarrow{PQ} = (6, -8)$
$\overrightarrow{QR} = (3, -4)$
$(\overrightarrow{PQ} - \overrightarrow{QR}) = (\_\_\_\_)$

Objective: Given $\overrightarrow{PQ} = (e, f)$ the student can correctly compute the length of a vector by applying the rule $m\ (\overline{PQ}) = \sqrt{e^2 + f^2}$

Items: Answer the following by applying the rule for finding the length of a vector. Show all work.

(a) 1. $\overrightarrow{PQ} = (3, -1)$ Find the length of $\overline{PQ}$
2. $-\overrightarrow{PQ} = (-3, 1)$ Find the length of $\overline{PQ}$

(b) $\overrightarrow{RS} = (-2, 4)$ Find the length of $3\ (\overrightarrow{RS})$

(c) $\vec{a} = (3, 2)$ Find the length of $\vec{a} + \vec{b}$
$\vec{b} = (-4, 3)$ Find the length of $\vec{a} - \vec{b}$

In triangle ABC
E is the mid point of $\overline{AB}$
F is the mid point of $\overline{BC}$
If $\overrightarrow{EF} = (3, -2)$
Find the length of $\overline{AC}$

Objective: Given that $\overrightarrow{PQ} = \overrightarrow{RS}$ and that $\overrightarrow{RS} = \overrightarrow{MN}$ the student can write $\overrightarrow{PQ} = \overrightarrow{MN}$ as a conclusion to the given.

Items:

(a) Given $\overrightarrow{PQ} = \overrightarrow{RS}$, $\overrightarrow{RS} = \overrightarrow{MN}$
If $\overrightarrow{PQ} = (3, -2)$ and $\overrightarrow{MN} = (e, f)$,
Find (e, f)

(b) Given $\overrightarrow{AB} = \frac{1}{2}\, \overrightarrow{CB}$
$\frac{1}{2}\, \overrightarrow{CB} = -\frac{1}{3}\, (\overrightarrow{PQ})$
If $\overrightarrow{PQ} = (6, -3)$ and $\overrightarrow{AB} = (e, f)$
Find (e, f)

(c) Given $\overrightarrow{AB} = -(1, \sqrt{3})$
$\overrightarrow{CB} = (-1, -\sqrt{3})$
$-\overrightarrow{BC} = \overrightarrow{AC}$
If $\overrightarrow{AC} = (e, f)$
Find (e, f)

Objective: Given a classroom development of a formal proof using vector notations and properties, the student can reproduce this proof in ten minutes or less, three days after the presentation. He can do at least eight of the proofs given in class. An example of the statement to be proven could be: For any parallelogram ABCD,

$$\overrightarrow{AB} = \overrightarrow{DC} \text{ and } \overrightarrow{AD} = \overrightarrow{BC}$$

Item: Write a formal proof for the following statements:

(a) For each line segment $\overline{PQ}$, if $\overrightarrow{PQ} = (e, f)$ and $e \neq 0$ then the slope of $\overline{PQ}$ equals $\frac{f}{e}$

(b) If a line joins the mid points of two sides of a triangle it is parallel to the third and equal to one half of it.

Objective: Given a set of statements relating to a vector diagram, the student can write at least one correct conclusion for the set of statements.

Item: In quadrilateral ABCD, $\overrightarrow{AB} \parallel \overrightarrow{CD}$

(a) Write at least one correct vector equation involving only the sides of the quadrilateral.

(b) Write at least one correct vector equation involving the diagonal CB

(c) Write at least one correct vector equation involving two adjacent sides and the diagonal CB

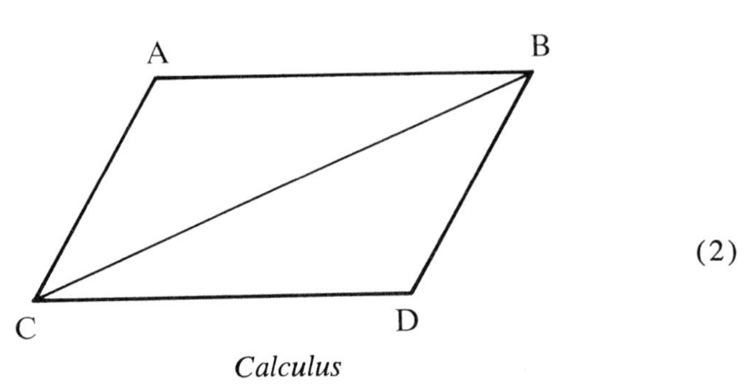

(2)

Calculus

The following objectives and related test items were written for a beginning calculus course.

Objective: Identify all the symbols used in first year calculus, e.g., $\frac{dy}{dx}$, f′, $\int f$, $\int_a^b f(x)dx$.

Item: Place one letter from Column B in each blank in Column A; each item of Column B may be used more than once.

Column A	*Column B*
__a__ $\frac{dy}{dx}$	(a) the instantaneous rate of change in y with respect to x
__a__ y′	(b) the differential of y
__e__ $\int_a^b f(x)dx$	(c) the limit of the function f(x) as x approaches a
__f__ $\frac{\Delta y}{\Delta x}$	(d) the sum of the values of f(x) at 1, 2, 3, –, –, –, n
__d__ $\sum_{k=1} f(x)$	(e) the definite integral
__b__ dy	(f) the slope of the chord line
	(g) the indefinite integral

Objective: Identify the geometric interpretation of the derivative and the definite integral, e.g., the derivative as the slope of the tangent line and the definite integral as the area under a curve.

Item: If f(x) is continuous on [a,b], then $\int_a^b f(x)dx$ may be interpreted geometrically as:

(a) the area under f(x) between a and b provided $f(x) > 0$.
(b) an approximation to the area under f(x) between a and b, provided a and b are both positive.
(c) the definite integral from a to b for any integers a and b.
(d) a rectangle whose base has length b − a and whose height is f(c) where c is in [a,b].
(e) none of these.

Item: Suppose that f(x) is defined by the graph:

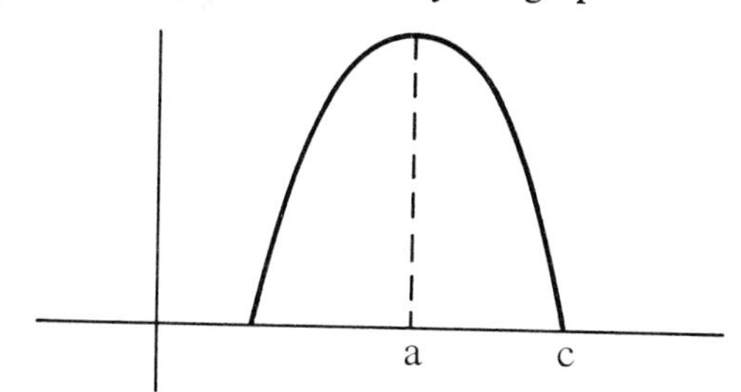

Which of the following is the best approximation to $\int_a^c f(x)dx$?

(a) $\frac{1}{2}$ (c − a) (f(a) − f(c)) (b) f(c) − f(a) (c) $\frac{f(c) - f(a)}{c - a}$

(d) $\frac{1}{2}$ (f(c) − f(a)) (e) $\frac{1}{2}$ $\frac{(f(c) - f(a))}{c - a)}$

Objective: Define intuitively the technical terms of first year calculus, e.g., the limit of a function, continuity, derived function, critical point, definite integral.

Item: Express in your own words the meaning of each of the following ideas; do not use such technical terms as limit, derivative, infinite series, antiderivative.

(a) The limit of a function at a point exists.
(b) A function is continuous at a point.
(c) The point (c,f(c)) is a critical point of the function f(x).
(d) The definite integral of a non-negative function exists for a given interval.
(e) The indefinite integral of a function exists.

Item: Each statement in Column B described a mathematical idea in rather loose non-technical terms. Match one item of Column A with each item of Column B; the items of Column A may be used more than once.

Column A	*Column B*
(1) f′(x) exists on the interval (a,b).	__4__ f(x) is neither rising nor falling.
(2) f(x) is continuous on the interval (a,b).	__1__ f(x) has a tangent line at every point between a and b.
(3) $\int_a^b f(x)dx$ exists.	__3__ The area under the graph of f(x) can be found if $f(x) > 0$ for $a < x < b$.

(4) $f'(c) = 0$.
(5) $\lim_{x \to c} f(x) = L$.

_____ f(x) is an unbroken curve.
_____ f(x) settles down as x gets close to c.
_____ The graph of f(x) may be traced without lifting the pencil.
_____ f(x) has no sharp corners.

Objective: Apply the geometric interpretation of the derivative and the integral.

Item: Draw the graph of a continuous function f(x) defined on [0,2] such that $f'(x) < 0$ when $x \neq 1$ and

(a) $f'(x)$ is continuous with $f'(1) = 0$.
(b) $f'(1)$ is undefined.
(c) $f'(x)$ is continuous and $f'(1) \to -1$ as $x \to 1$.

[Note: Draw a separate graph for each of parts a,b, and c.]

Item: Let $a < b < c < d$ be elements of the domain of a continuous function f. Sketch the graph of the function if all of the following conditions hold:

(a) $\int_b^c f(x)dx + \int_c^d f(x)dx > 0$
(b) $\int_c^d f(x)dx < 0$
(c) $\int_a^b f(x)dx = \int_b^c f(x)dx$
(d) $\int_a^c f(x)dx > 0$

Objective: Recognize the importance of the fundamental theorem of calculus in relating differentiation and integration.

Item: The fundamental theorem of calculus states

$$\int_a^b f(x)dx = F(b) - F(a).$$

The importance of this theorem lies in the fact that it relates the basic concepts of

(a) function and integration
(b) limit and integration
(c) continuity and integration
(d) differentiation and integration
(e) slope and integration

Item: Choose from the list below the statement which best relates the concept of differentiation to integration.

(a) $\int_a^b f(x)dx = \lim_{\triangle x \to 0} \sum f(x^*) \triangle x$

(b) $\int f(x)dx = F(x) + c$ where $F'(x) = f(x)$

(c) $D_x(\int f(x)dx) = f(x)$
((d)) $\int_a^b f(x)dx = F(b) - F(a)$ where $F'(x) = f(x)$
(e) There is no relationship between differentiation and integration.
(7)

Elementary School Science

The following is a portion of a UNIPAC for a simple and compound machines unit. The suggestions for evaluation for teachers follow the unit. See also the sample questions listed in the pictorial riddle and junior high school section of chapter eight. Many of them have relevance for the elementary school.

STUDENT'S SECTION

There's a Simple Machine in Your Life

Learning Objectives

1. To discover the mechanical advantage of the pulley.
   You will be able to construct a pulley system and compare (measure) the amount of force needed with and without the pulley.
2. To discover the uses and advantages of the lever, the inclined plane, the wedge, the wheel, and axle.
   You will be able to demonstrate in any fashion you consider suitable, the use of at least five different applications of simple machines which are used in and around your home.
3. To tell how simple machines are put together to make compound machines.
   You will be able to describe, demonstrate the use of, or construct at least one compound machine which is used in your home.

Pre-Evaluation

Look at the learning objectives. Have you read them carefully and do you understand them? If you understand them and think you are interested in learning more about simple and compound machines, turn to the next page.

LESSON TOPIC: *Simple Machines I*

I. *Objective:*

   To discover the mechanical advantage of the pulley.
   You will be able to construct a pulley system and compare (measure) the amount of force needed with and without the pulleys.

II. *Activities:*

A. Laboratory—The materials listed below are to help you do an activity related to your objective. Do any or all of the activities or one of your own choosing. You determine how to use the material to reach your objective.

25–1. Use of a pulley
Spring balance, some string, one pulley, and a weight

25–2. Using a block and tackle.
Spring balance, two single and two double pulleys
Several feet of string and some weights

In the above activities use the spring balance to determine how much effort is required without using the pulley and how much is required with the pulley.

B. Identify below the sources of information that you use.

1. Topics to assist you in locating information: (1) Simple Machines, (2) Pulley ____________________
2. Films: ____________________
3. Filmstrips: ____________________
4. Tapes: ____________________
5. Textbooks: ____________________
6. Books: ____________________
7. Vertical File: ____________________
8. Field Trip: ____________________
9. People: ____________________
10. Magazines: ____________________
11. Other: ____________________

III. *Self-Evaluation:*

A. What have you learned?

B. Can you answer any of these questions?

How does the use of a pulley or pulleys make work easier?

Can you lift more weight with a block and tackle than with a single pulley?

What is the difference betweeen a fixed pulley and a moveable pulley?

IV. *Self-Evaluation Key:*

See your teacher for a copy of his answers to the above questions so that you can compare and discuss any differences between your answers and his.

V. Questions:

Do you want to do further study on this subject?

Do you have any questions that you want to investigate further?

LESSON TOPIC: *Simple Machines II*

I. *Objective:*

To discover the uses and advantages of the lever, the inclined plane, the wedge, and the wheel and axle.

You will be able to demonstrate in any fashion you consider suitable, the use of at least five different applications of simple machines which are used in and around your home.

II. *Activities:*

A. Laboratory — The materials listed below are to help you do an activity related to your objective. Do any or all of the activities or one of your own choosing. You determine how to use the material to reach your objective.

26–1. Locating the fulcrum.
A weight, a ruler, a fulcrum and a spring balance.

26–2. Using the lever.
A weight, string, a spring balance and a ruler.

26–3. The screw and the wedge are a form of inclined plane.
You find a way to show the advantage of using these three simple machines.

26–4. Prepare a list of uses that the wheel and axle can be put to.

B. Identify below the sources of information that you use.

1. Topics to assist you in locating information: (1) Simple Machines, (2) Levers, (3) Incline Planes, (4) Wheel and Axle
2. Films: ____________________
3. Filmstrips: ____________________
4. Tapes: ____________________
5. Textbooks: ____________________
6. Books: ____________________
7. Vertical File: ____________________
8. Field Trip: ____________________
9. People: ____________________
10. Magazines: ____________________
11. Other: ____________________

III. *Self-Evaluation:*

A. What have you learned?

B. Can you answer any of these questions?
How does the position of the fulcrum make work easier?

Can you explain how the screw is considered to be a form of the inclined plane?

You have two bikes, one with large wheels and one with small wheels. If you apply the same amount of effort, the large wheel bike will go faster. Why?

IV. *Self-Evaluation Key:*

See your teacher for a copy of his answers to the above questions so that you can compare and discuss any differences between your answers and his.

V. Questions:

Do you want to do further study on this subject?

Do you have any questions that you want to investigate further?

LESSON TOPIC: *Simple Machines III*

I. *Objective:*

To tell how simple machines are put together to make compound machines.

You will be able to describe, demonstrate the use of, or construct at least one compound machine which is used in your home.

II. *Activities:*

A. Laboratory — The materials listed below are to help you do an activity related to your objective. Do any or all of the activities or one of your own choosing. You determine how to use the material to reach your objective.

27–1. Prepare a list of compound machines and point out the simple machines that make up the compound machine.

27–2. Draw pictures of five compound machines and point out with arrows the simple machines being used.

Remember, a compound machine is one that is made of more than one simple machine.

B. Identify below the sources of information that you use.

1. Topics to assist you in locating information:
(1) Machines, (2) Compound Machines __________
2. Films: __________
3. Filmstrips: __________
4. Tapes: __________
5. Textbooks: __________
6. Books: __________
7. Vertical File: __________
8. Field Trip: __________
9. People: __________

10. Magazines: ______________________________
11. Other: ______________________________

III. *Self-Evaluation:*

A. What have you learned?
B. Can you answer any of these questions?

Do you have any compound machines around your house? Name some of them.

Is your bike a compound machine? Why, or why not?

Is a gear a simple or compound machine?

IV. *Self-Evaluation Key:*

See your teacher for a copy of his answers to the above questions so that you can compare and discuss any differences between your answers and his.

V. Questions:

Do you want to do further study on this subject?

Do you have any questions that you want to investigate further?

TEACHER'S SECTION

Instructions and Evaluation

I. *Pre-Test and Pre-Test Keys:*

You will note that the learning objectives for this UNIPAC describe behaviors that can be evaluated only en route. Therefore, pre- and post-testing is impossible. However, pre-evaluation of a type does occur because the student bases his selection of this UNIPAC on his own interest, Post-evaluation, based on the learning objectives, is the responsibility of the teacher and is based on observation, discussions with the student, and/or written reports from the student.

II. *Special Instructions for the Self-Test:*

You will find answers for the self-tests below. Please note that the self-test questions *do not ask* for behaviors directly related to the learning objectives because the learning objectives call for accomplishment of en route behaviors. The self-tests are intended to help the student check his own transfer of knowledge to new situations. However, *teacher evaluation is* based on the learning objectives which have been given to the student.

A. Self-Test Key — Simple Machines I

1. A pulley makes work easier by changing the direction of force. Also, by increasing the number of ropes and pulleys used, you can decrease the amount of force needed to move a load.

2. Yes, you can lift more weight with a block and tackle than with a single pulley because you have increased the number of supporting ropes used and thereby have multiplied your force.
3. The fixed pulley is fastened to a fixed support whereas the movable pulley is not. The fixed pulley is more convenient because it changes the direction of force. However, the movable pulley will cut the effort in half and so multiplies the force by two.

B. Self-Test Key — Simple Machines II

1. The closer the fulcrum is to the object to be moved the easier it will be to move it.
2. If you examine a screw closely you will discover it is just a winding incline plane that comes to a point.
3. The bike with large wheels will go faster than one with small wheels if the same amount of effort is applied because the large wheeled bike will go farther each time the wheel goes around. (It covers more distance with each revolution.)

C. Self-Test Key — Simple Machines III

1. Any combination of two or more simple machines is a compound machine. There are many in each home — check the ones you named with your teacher.
2. Yes, your bike is a compound machine because it is made of two or more simple machines.
3. A gear is a simple machine. It is an adaptation of the wheel and axle. (4)

Sample Objectives and Test Questions For Junior High School

The following are examples of junior high school objectives and tests written for them.

Ninth Grade Science Objectives Organized According to Bloom's Taxonomy

The High School Entrance Examinations Board, Department of Education, Edmonton, Alberta, published a booklet containing sample objectives and test items for ninth grade science classified according to Bloom's *Taxonomy*. Listed below are sections applied only to the Application through Evaluation levels of *The Taxonomy*.

3. Application

Application can best be described by comparing it with comprehension. A problem in the comprehension category requires the student to know an abstraction well enough that he can correctly demonstrate its use when he is specifically asked to do so. Application, on the other hand, requires a step beyond this. Given a problem new to the student, he will apply the appropriate abstraction without having to be prompted as to which abstraction is correct or without having to be shown how to use it in that situation. In comprehension the student shows that he *can use* the abstraction when its use is specified. In application the student shows that he *will use* the abstraction correctly given an appropriate situation in which no mode of solution is specified.

It is not necessary always to require a complete solution. Sometimes a partial solution or selection of the correct abstraction alone is requested. In all cases, however, the task should use material the student has not had contact with, or take a new slant on common material or use a fictional situation. If the material is familiar, comprehension or recall is required, not application.

Some objectives

1. Applies scientific concepts and principles used in class to the phenomena discussed in a paper.
2. Predicts the probable effect of a change in a factor on a chemical solution previously at equilibrium.
3. Applies scientific principles, postulates and other abstractions to new situations.
4. Finds solutions to problems in making home repairs employing experimental procedures from science.

Some Items

1. What effect would an increase in pressure in the boiler of a steam engine have upon the performance of the steam engine?
2. A baseball can be made to curve. How can this be explained?
3. Would the air in a closed room be heated or cooled by the operation of an electric refrigerator in the room with the refrigerator door open?

   *A. Heated, because the heat given off by the motor and the compressed gas would exceed the heat absorbed.
   B. Cooled, because the refrigerator is a cooling device.
   C. Cooled, because compressed gases expand in the refrigerator.
   D. Cooled, because liquids absorb heat when they evaporate.
   E. Neither heated nor cooled.

• • • • • •

Items 4 and 5 are based upon the following situation.

The boiling and freezing points of water were determined and marked on the glass of a new, and as yet "blank" thermometer. If these two points were 9 inches apart, how far apart would the degree markings be if it were desired to make

4. A Centigrade thermometer?
   A. 9/180 inch
   *B. 0.09 inch

C. 0.9 inch
D. 9/16 inch
E. 9/32 inch

5. A Fahrenheit thermometer?
*A. 0.05 inch
B. 0.5 inch
C. 0.09 inch
D. 9/100 inch
E. 9/32 inch

• • • • • • •

6. An automobile weighing 3,300 lbs. goes up a hill 160 ft. high in 20 secs. The output of the motor in horsepower must be
A. 8
*B. 48
C. 412.5
D. 10,560,000

Note: Item 6 should be classed as "Comprehension" if it is parallel to, or similar to, items taken in class. In Items 4 and 5 the situation is probably novel — it has a new twist. Item 6 might be made novel to most students, for example, by including the length of the hill.

4. *Analysis*

Analysis is related to both comprehension and evaluation. In comprehension the emphasis is on the grasp of the meaning and intent of the material. Analysis emphasizes the breakdown of the material into its constituent parts and detection of the relationships of the parts and of the way they are organized. It is also sometimes directed at the techniques and devices used to convey the meaning or to establish the conclusion of the communication. Comprehension deals with the content of the material alone, analysis with both content and form. On the other hand, evaluation involves more than analysis or even critical analysis in that, when one evaluates, in addition to analyzing a judgment is made in terms of some criteria as to its adequacy.

Some Objectives

1. Recognizes unstated assumptions.
2. Distinguishes facts from hypotheses.
3. Distinguishes conclusions from statements which support it.
4. Checks consistency of hypotheses with given information and assumptions.
5. Distinguishes cause and effect relationships from other sequential relationships.
6. Detects the purpose, point of view, attitude, or general conception of the author.

Some Items

1. "Light is produced indirectly in the fluorescent lamp." What would be another way of expressing the conclusion implied in the above statement?
2. 1 liter = .88 qt. Are there more quarts than liters in a container whose capacity is 1000 cc?

• • • • • •

Item 3 contains a pair of statements which are either in agreement with each other or not in agreement, and either of the statements may be true or false. Study the pair of statements and choose

A. If statements I and II are in agreement and both false.
B. If statements I and II are in agreement and both true.
C. If statements I and II are not in agreement; I true, II false.
D. If statements I and II are not in agreement; I false, II true.

3. I. At absolute zero the molecules of a substance do not move with respect to each other.
   II. No heat energy is possessed by a substance at absolute zero.

   (Key: (B) *Note:* This pair of statements is only illustrative as other pairs of statements could be added to make use of the same alternatives.)

• • • • • •

The following paragraphs concern the action of a geyser. Read the passage carefully before answering Items 4-8 inclusive.

A geyser is a hot spring that erupts at intervals. It is made up of a more or less crooked or constricted tubular fissure that extends into the earth and is filled with water. A source of heat near the bottom of the fissure heats the water.

After an eruption the tube fills with water from an underground source. The water throughout most of the length of the tube, and especially in the lower part, becomes heated to a point above the normal boiling temperature (212°F.) of water but does not become quite hot enough to turn to steam. However, sooner or later, some of the water in the lower part of the tube at the source of heat reaches the boiling point and turns to steam. The steam raises the whole column of water above it and causes some water to overflow from the geyser pool at the surface. This overflow acts as a trigger, permitting the whole column of water in the tube to flash into steam which blows from the fissure in an eruption.

For each of Items 4-6 choose

A. If the statement is true and pertains directly to the action of the geyser.
B. If the statement is true but is not directly concerned with the action of the geyser.

C. If the statement is false.

4. If the tube were not crooked and constricted the water throughout the tube would come to nearly the same temperature by unrestricted convectional circulation, the water would boil, and a boiling spring rather than a geyser would result.
5. The boiling point of water in the bottom of the tube is lower than it is at the top.
6. The water in the tube does not turn to steam although it is above the normal boiling point because of the pressure of the overlying water.

(Keyed answers are: 4 (A), 5 (C), 6 (A).)

• • • • • •

7. Air pressure on a fine day is usually higher than on a stormy day. Thus, the geyser will erupt more often during stormy weather. This statement can best be evaluated as
   A. A wild guess.
   *B. True, because it can be predicted from known principles.
   C. False, because it cannot be predicted from known principles.
   D. True, because it is based on statements in the above passage.
   E. False, because it contradicts the statements in the above passage.
8. An investigation of which one of the following hypotheses concerning the boiling point of water is most suggested by the observation of a geyser?
   A. The boiling point of water decreases with increasing altitudes.
   B. The boiling point of water is that point when the vapor pressure equals the atmospheric pressure.
   *C. The boiling point of water in a pressure cooker is above 212°F.
   D. The boiling point of water in a hot water heating system is above 212°F.

(*Note:* Item 4. is classified as Analysis, 5. as Knowledge of Specifics, 6. as Comprehension, 7. as Evaluation, and 8. as Synthesis.)

5. Synthesis

Synthesis emphasizes the putting together of elements and parts in such a way as to constitute a pattern or structure not clearly there before. Generally this involves a recombination of parts of previous experience with new material reconstructed into a new and more or less well-integrated whole. This is the category in the cognitive domain which may not always be free creative expression since generally the student is expected to work within the limits set by particular problems, materials, or some theoretical or methodological framework.

Some Objectives

1. Describes an original experiment conducted by a student.
2. Proposes ways of testing an hypothesis.

3. Integrates the results of an investigation into an effective plan or solution to solve a problem.
4. Develops schemes for classifying chemicals.
5. Formulates hypotheses to explain adequately a wide range of seemingly interrelated phenomena and be internally consistent.
6. Makes deductions from basic theories.

Some Items[5]

1. How could you construct an oil barometer from oil which floats on water?
2. Given: 1. Work = fd.
   2. Power is work per unit of time.
   3. 33,000 ft. lb. per min. = 1 H.P.

   Develop the formula for Horsepower H.P. =
3. You suspect that a certain soluble compound contains hydrogen. Which of the following would be a way of checking your suspicion?
   A. Prepare a solution of the compound and subject it to electrolysis, using Hoffman's apparatus.
   B. Place a sample of the compound in a Bunsen flame and determine if any water vapor is produced.
   C. Heat the compound to incandescence and analyze the light produced with a pyrometer.
   *D. Vaporize some of the material in a flame and use a spectroscope to analyze the light produced.
   E. Pass some light through a solution of the compound and analyze the spectrum produced with a spectroscope.
4. See also Item 8 under Analysis.

6. Evaluation

Evaluation includes the process of making judgments concerning the extent to which ideas, solutions, methods, and material satisfy criteria. It involves the use of criteria as well as standards for appraising the extent to which particulars are accurate, effective, economical or satisfying. The judgments may be either quantitative or qualitative, and the criteria may be either those determined by the student or those which are given to him.

Some Objectives

1. Detects the logical fallacies in arguments.
2. Discerns the inconsistencies and/or logical inaccuracies in data.
3. Identifies the procedural errors in an experiment and evaluates their effect on the results.
4. Compares the effectiveness of a method used to solve a problem with the best possible method available.

---

[5]It is extremely difficult to write objective items to measure synthesis and evaluation. Bloom provides several examples in science. The Physical Science Study Committee tests published by Educational Testing Service also have examples.

5. Identifies additional evidence or experimentation that is necessary to fully justify the conclusions reached.

Some Items

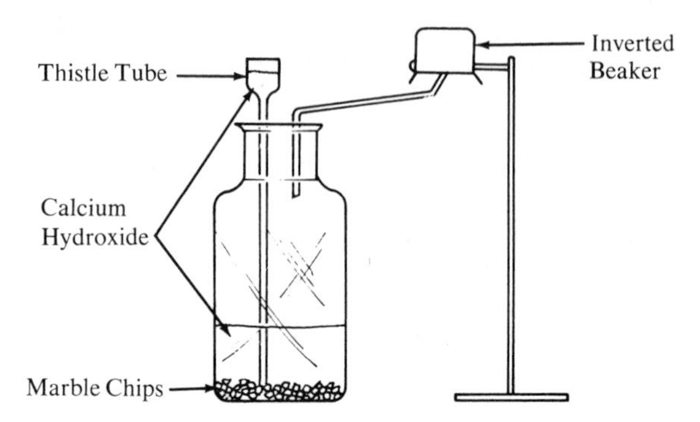

The above apparatus was set up to prepare carbon dioxide. The contents of the inverted beaker were tested with lime water which did not turn milky. It was concluded that the calcium hydroxide solution was not strong enough.

1. What logical inconsistencies are there in the report of the experiment?
2. Identify the procedural errors in the experiment and evaluate their effect on the results.
3. A recent popular article stated that, if a space craft were being rocketted away from the surface of the earth fast enough, its speed would offset the pull of gravity and a man in the rocket would feel weightless. This statement should be evaluated as being
   - A. Accurate since two forces are involved in the described situation and two forces can offset one another.
   - B. Accurate since the weight of an object decreases rapidly as the distance from the center of the earth decreases.
   - *C. Inaccurate since the floor of the man's compartment will be pushing up on the man as the rocket rises, and this push will give a feeling of increased rather than decreased weight.
   - D. Inaccurate since no object within the gravitational field of the earth can be weightless and so cannot seem weightless.
   - E. Inaccurate since the chance that the force on the man due to the rocket moving up and the force of gravity pulling the man down will be equal is extremely minute (9, pp. 17-24).

ESCP

The following are examples of test questions written for the Earth Science Curriculum Project. Note that the instructor preparing these indicated in the margin for each objective its cognitive level. Note also that the test involves the use of several audio-visual aids.

Knowledge:

1. Identify a sequence of layered rock as sedimentary rock when shown a slide, picture, or transparency of several layers of rock.

Part I: The following multiple choice questions are to be answered after observing slides. Circle the correct answer for each question.

1. The students will be shown a slide of some obvious beds of sedimentary rock such as is exposed in the Badlands of South Dakota.
   Question: What class of rock is shown in this slide?
   a. Igneous Intrusion
   b. Sedimentary
   c. Metamorphic
   d. Igneous Extrusion
   e. Volcanic Lava Flow

Application: 2. Make an inference as to why several layers of sedimentary differ in appearance when shown a slide, picture, or transparency of some distinguishable rock layers.

2. The students will be shown a slide of some tilted strata.

   Question: Which statement would be true about the history of the region shown in the slide?
   a. Deposition of beds on a mountain slope.
   b. Deposition of beds followed by the tilting of the beds.
   c. Several lava flows have flowed over this region.
   d. Several igneous intrusions have been exposed by weathering.
   e. None of these statements could be true.

Application: 3. State a brief geological history when shown a slide, picture, or transparency of some tilted sedimentary layers.

3. The students will be shown a slide of some cross-bedding.

   Which statement would be true about the cause of the geological feature shown on this slide?
   a. Glaciers, passing over some sandstone, have carved this region.
   b. The continuous shifting and depositing of sand dunes.
   c. Running water has dissolved thin layers of rock.
   d. Tremendous pressure and heat have caused layers to form.
   e. Earthquakes have caused the layers to fold.

Application: Correlate some rock layers across a fault when shown a cross section of a fault which indicates rock types.

4. The students will be shown a slide of a fault. This fault should show an obvious contrast in color or an obvious displacement of a conspicuous layer.

   What geological feature is shown in this slide?
   a. Cross-bedding
   b. Fault

c. Igneous Intrusion
d. Foliation
e. Jointing

Analysis: Make an inference of a brief history of a rock when given an sample of breccia, well-rounded sandstone, poorly-rounded sandstone, or conglomerate.

5. The students will be shown a slide of some cross-breeding.

   What geological feature is shown in this slide?

   a. Cross-bedding
   b. Fault
   c. Igneous Intrusion
   d. Foliation
   e. Jointing

Synthesis: Reconstruct the sequence of deposition when given a cross section of several beds which includes an angular unconformity, and a lava flow.

6. The students will be shown a slide of some obvious beds of sedimentary rock.

   These layers differ in appearance. Which statement below could explain this difference in appearance?

   a. Differ in mineral content
   b. Differ in density
   c. Differ in size of sediment
   d. Differ in age of sediment
   e. Both a. and c. are true (10)

Science in the High School

The following are examples of objectives and tests written for them for the various school courses.*

Sample Objectives and Test Questions for High School Science

Biology

Listed below are sample objectives and test questions for the Second Edition of BSCS Green Version Biology, chapter 14.

The student should be able to:

*No examples are given for Harvard Project Physics (11) because unit tests are provided by this curriculum project and have been designed around behavioral objectives. These tests are available from the publishers of the materials.

Objective 1: Compare at least eight ways that the external structure of the frog and man are similar and different.

1. Put an "S" by the structure of the frog listed below if it is similar to man or a "D" if dissimilar.
   ______a. Neck
   ______b. Trunk
   ______c. Eyes
   ______d. Eyelids
   ______e. Ears
   ______f. Structures on fingers
   ______g. Body symmetry
   ______h. Appendages

Objective 2: Associate organs and chemicals with the physical and chemical phases of digestion.

2. Place a "P" by the phrase if it belongs to the physical stage of digestion and a "C" if it is related to the chemical stage.
   ______a. Chicken gizzard
   ______b. Lion's teeth
   ______c. Steak in your stomach
   ______d. Action of the esophagus
   ______e. Chyme
   ______f. Bile
   ______g. Amino acids

Objective 3: Develop an hypothesis concerning the efficiency of the absorption of the villi and alveoli using the information given in this chapter.

3. The efficiency of the alveoli and villi can be attributed to
   a. The oxygen content of the lungs
   b. The smallness of the structures
   c. The amount of surface covered by these structures
   d. The absorbing potential of each structure
   e. The relationship found between the two structures

Objective 4: Identify the location of a physical and chemical process when given a list of food which has been consumed.

4. When eating french fried potatoes, chemical digestion begins in:
   a. The mouth
   b. The esophagus
   c. The small intestine
   d. The stomach
   e. The large intestine

Objective 5: Trace the circulation of the blood through the body integrating it with the respiratory system when given a model of the heart (or an actual specimen). The function of the valves of the heart should be included.

5. When a baby is born sometimes he is referred to as a "blue baby". This occurs when there is an opening between the right and left side of the heart. The reason for the descriptive word "blue" might be:
   a. All oxygenated blood is shown in blue in scientific drawings
   b. All deoxygenated blood is shown in blue in scientific drawings
   c. Some of the blood does not have a chance to return to the lungs and is mixed with oxygenated blood and the baby turns blue from lack of oxygen.
   d. The baby turns blue because he has received too much oxygenated blood
   e. Some of the blood returning from the lungs mixed with the blood in the right side of the heart.

Objective 6: Infer what is possibly occurring when given a problem stating that a small child has small knots on his neck.

6. A small child is feeling fine, yet when his mother washed behind his ears, she noted he had "lumps" behind them. He might have
   a. The mumps
   b. Received a blow behind the ears in two places
   c. An infection causing the lymph nodes to swell
   d. A growth under the skin on the mastoid bone
   e. An accumulation of red blood cells blocking a vein or artery

Objective 7: Hypothesize what might happen to a herbiverous vertebrate if the cellulose-digesting microorganisms were not present in its digestive tract.

7. Which of the following graphs would show what might happen if the cellulose-digesting microorganism were not present in a cow?

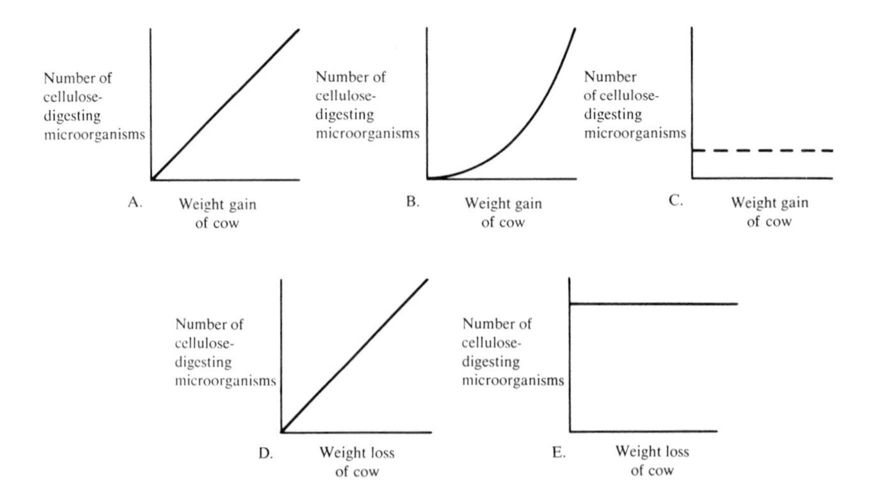

Objective 8: State an appropriate hypothesis concerning the effects of varying environmental temperature on the heartbeat of a small animal after completing Investigation 14.3 and with the use of data from the laboratory work.

8. According to data collected in Investigation 14.3, the following hypothesis may be stated:
   a. The higher the temperature of the water, the faster the Daphnia's heart beats.
   b. The higher the temperature of the water, the slower the Daphnia's heart beats.
   c. The lower the temperature of the water, the faster the Daphnia's heart beats.
   d. The higher the temperature of the water, the slower the Daphnia's heart beats.
   e. You can obtain the average heart beat of the Daphnia by placing dots on paper for fifteen seconds every time you observe a heart beat, then multiply by four.

Objective 9: List at least six structures of the respiratory system and explain their function when given an unlabeled diagram of the system.

9. Listed below are six structures of the respiratory system with their functions. Place a "T" if the structure and listed function is correct, and an "F" if the relationship is false.
   ______a. Ribs—protection of the lungs
   ______b. Nasal cavity—warms and moistens the air
   ______c. Trachea—carries air to the lungs
   ______d. Alveoli—allows for diffusion of carbon dioxide and oxygen
   ______e. Epiglottis—keeps food out of the trachea
   ______f. Diaphragm—regulates air pressure inside the lungs

Objective 10: Identify William Harvey and Marcello Malpighi by relating their contributions to human physiology by doing library research supplementing text materials.

10. Because of the work done by William Harvey and Marcello Malpighi, people living in the 17th Century had knowledge of
    a. The reproductive system
    b. That mammals have four-chambered hearts
    c. That a sperm cell has a flagellum
    d. Arteries, veins, and capillaries
    e. The diffusion of gases that takes place between the capillaries and tissues

Objective 11: Make a chart comparing the circulatory systems of man with lower organisms, including at least five categories. The text may be used in making this chart.

11. In comparing man's circulatory system with that of lower animals, which of the following statements is incorrect.

a. Both the clam and man have closed circulatory systems
b. Both man and the earthworm have closed circulatory systems
c. All mammals receive blood from the body in the right side of the heart
d. The circulatory system of man and lower animals both serve the same function
e. In the fish, frog and man there are valves that keep the blood from going back and forth between the heart chambers

Objective 12: Design and carry out an experiment to see what would happen to the human heart beat if there were changes in the environment temperature using the results obtained in Investigation 14.3.

12. Which of the following graphs might represent the human heart beat in relation to environmental temperature?

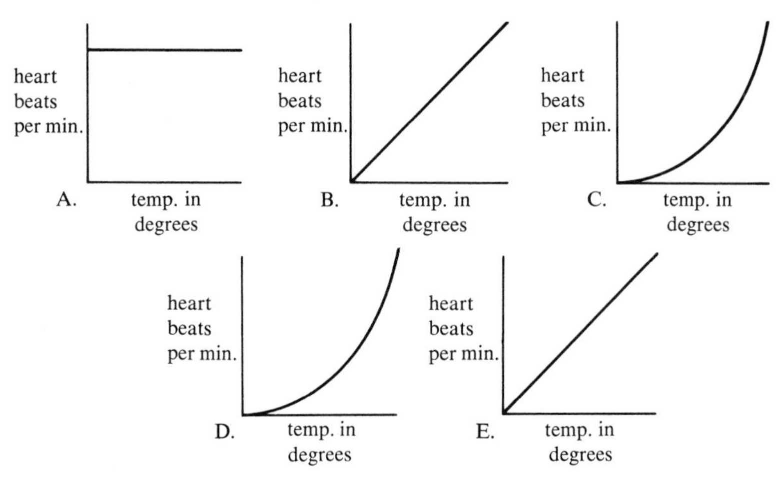

Objective 13: Distinguish between excretion, secretion, and elimination when given specific examples of these processes.

13. Place an "EX" if the example listed below is a form of excretion; "EL" if a form of elimination; or "S" if a form of secretion.
_____a. Nasal drip
_____b. Releasing of a hormone
_____c. Saliva
_____d. Undigested cells
_____e. Pus
_____f. Diffusion of carbon dioxide from the cell
_____g. Thrombokinase
_____h. Urine

Objective 14: Hypothesize the work of the kidney when given certain situational problems. For example, that which will happen to the water content in a person's blood if he attends a game in 100 degree weather and drinks three bottles of pop?

14. While Joe attended a baseball game with the temperature reading at 75 degrees, he drank four bottles of pop. The water content of his blood will probably be
    a. Higher than normal
    b. About the same as normal
    c. Lower than normal
    d. High because of little perspiration
    e. Low because of urine

Objective 15: Name the gland when given a list of glandular disabilities that is involved with the disorder. The student may use a list he has compiled himself.

15. Using your notes, match the proper gland with the physical characteristic listed below. The gland may be used more than once.

______a. High blood pressure	1. Thyroid gland
______b. Appearance of whiskers	2. Islands of Langerhans
______c. Over-weight	3. Parathyroid gland
______d. Diabetes	4. Adrenal glands
______e. Excess energy	5. Testes
______f. Faulty diaphragm movement	

Objective 16: Classify when given a list of at least ten voluntary and involuntary actions and describe the differences in these processes occurring in the nervous system.

16. Place an "I" by the action if it is involuntary, and a "V" if it is a voluntary action.
    ______a. Typing
    ______b. Flinching
    ______c. Catching a ball
    ______d. Winking at the opposite sex
    ______e. Production of perspiration
    ______f. Blushing
    ______g. Combing your hair
    ______h. Blowing your nose
    ______i. Crying
    ______j. Talking

Objective 17: Classify muscles as smooth or striated when given a list of muscles.

17. Write the term striated or smooth by the muscles listed below.
    ______________a. Diaphragm
    ______________b. Stomach
    ______________c. Biceps

_______________d. Steak
_______________e. Walls of the large intestine

Objective 18: Substantiate or disprove the idea that man is the highest form of life from what you have learned about the various systems of the human body in this chapter.

18. Write a short essay substantiating or disproving the idea that man is the highest form of life.

Objective 19: The student should be more curious for current information than is in the text concerning the human body, so that when given an opportunity to do extra credit work by watching a given television program the student will do so.

19. If given an opportunity check two of the items listed below which you would like to do.
    _____a. Watch a TV program dealing with the human body.
    _____b. Read a recent article in a magazine on the human body.
    _____c. Do another experiment to discover a new concept.
    _____d. Design an experiment.
    _____e. Visit a blood bank and observe procedures.
    _____f. None of the above

Objective 20: Appreciate the intricacies of the human body enough to contact groups like the Heart Association, Leukemia Association, Cancer Society, county health departments, and other related organizations for additional material not only for himself, but for the entire class.

20. Place a check by any of the agencies you have recently and voluntarily contacted in order to obtain more information about the human body and disease.
    _____ Heart Association
    _____ Muscular Dystrophy Association
    _____ Cancer Society
    _____ Leukemia Association
    _____ Tuberculosis Association
    _____ Local blood banks
    _____ County health agencies
    _____ Polio agency
    _____ Others (Name these) (3)

PSSC Chapter 5

The following are examples of objectives and test questions for the Physical Science Study Committee physics book. In each of the following the objective is given first and then the test question to determine its attainment. After studying chapter five, a student should be able to:

Objective 1: Describe the position of an object in terms of its "x" coordinate, if given a specific origin and a directed axis.

Suppose we choose the direction east to be positive and west to be negative. A car leaves the intersection of Main and Central avenues traveling east, and stops after traveling four blocks.

a. What is the position in terms of the origin?
b. If the car had traveled west a similar distance, what would its position be?

Objective 2: Define displacement, velocity, and acceleration in terms of the "x" coordinate and the time.

Match the correct definition with the appropriate term.

______Velocity	A. $\triangle$ x
______Acceleration	B. $\triangle$v/$\triangle$ t
______Displacement	C. $\triangle$ x/$\triangle$ t

Objective 3: Construct a graph of position versus time if given a series of displacements corresponding to a series of times.

An olympic runner decides to train on a straight stretch of highway. Below is a table of his position and corresponding times. Plot a graph of his position versus time.

Position	*Time*
1. + 4 miles	0 minutes
2. + 2 miles	10 minutes
3. − 2 miles	35 minutes
4. 0 miles	50 minutes
5. + 6 miles	110 minutes

Objective 4: Construct a graph of velocity versus time if given a series of displacements corresponding to a series of times.

Here are some displacements each of which occurred in consecutive time intervals of 0.4 seconds. Plot the velocity-versus-time graph for this data.

$d_1 = +2$ cm	$d_5 = +10$ cm
$d_2 = +3$ cm	$d_6 = 0$ cm
$d_4 = -3$ cm	

Objective 5: Construct a graph of acceleration versus time if given a series of velocity changes corresponding to a series of times.

Sketch a graph which shows the acceleration of a motorcycle whose velocity at a time, t = 0 is Θ mph, and at a time, t = 5 seconds its velocity is 60 mph. How would your graph look if the motorcycle had been slowing down?

Objective 6: Calculate the velocity of an obiect if given a graph of its position versus time.

Below are three position versus time graphs. For each, sketch the graph of velocity versus time.

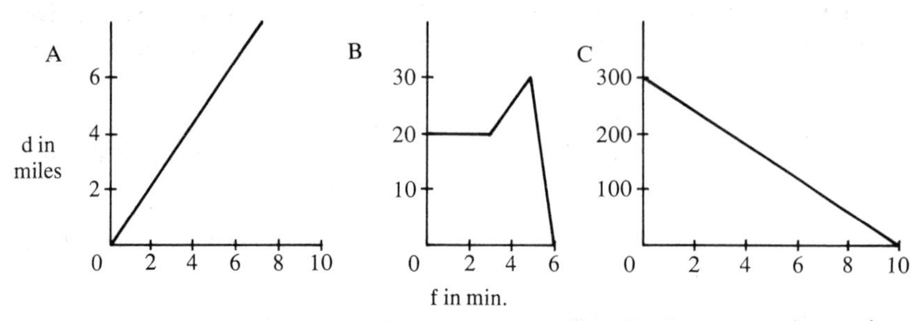

Objective 7: Calculate the acceleration of an object in the appropriate units, if given a graph of its velocity versus time.

What conclusions can you make concerning the acceleration of the objects whose velocity versus time graphs are shown below?

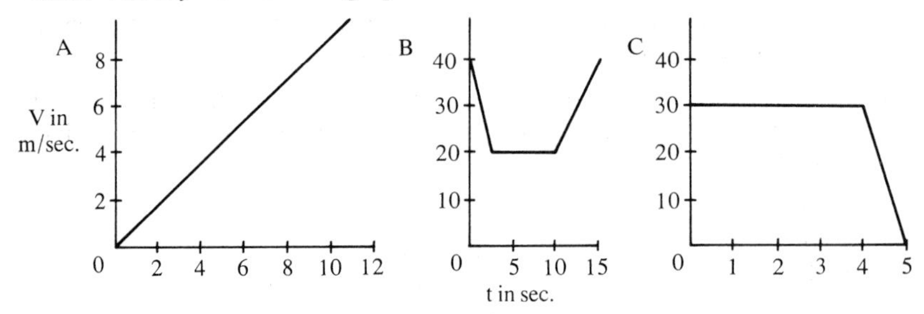

Objective 8: Calculate the displacement of an object if given a graph of its velocity versus time.

Calculate the displacement of the object in each of the three graphs of question 7.

Objective 9: Analyze any one of the three types of graphs and construct the other two graphs from this analysis.

From this graph of velocity versus time for a train starting from rest, plot graphs of position versus time, and acceleration versus time.

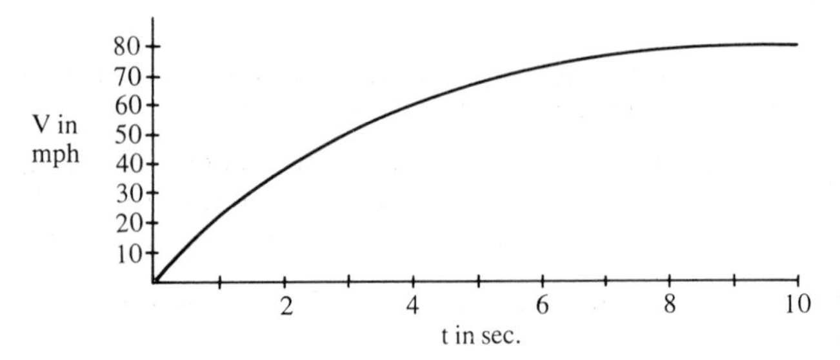

Objective 10: Define "slope" in terms of the coordinates of the vertical and horizontal axes.

a. Define "slope" in terms of the "x" and "y" axes.
b. What does the slope of a position-time graph tell you?
c. What does the slope of a velocity-time graph tell you?

Objective 11: Contrast "average"and "instantaneous" when these terms are applied to velocity or acceleration.

Below is a position versus time graph.

a. How would you calculate the average velocity for the time interval?
b. How would you calculate the instantaneous velocity, and how does it compare with the average velocity?

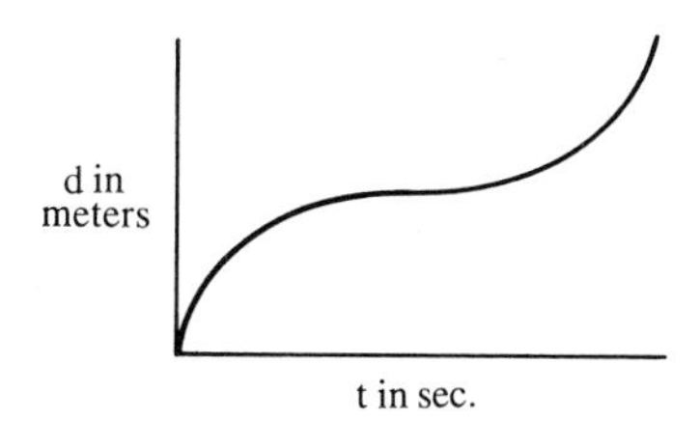

Objective 12: Derive the four motion equations if given a graph of the velocity versus the time.

a. How was this equation, $v_f - v_i = a(t_f - t_i)$, derived?
b. Which area formula from geometry does this equation remind you of, $d = ½ (v_f + v_i) t$? Sketch a graph showing its origin.
c. Refer to the equation in part b. If $t = (v_f - v_i)/a$, solve the final velocity.
d. Solve the equation in part a. for the final velocity. Substitute that expression into the equation in part b and derive an expression for the displacement, d.

Objective 13: Solve any given problem concerning motion along a straight line using the motion equations.

Using the motion equations you have derived solve the following problem. A bobsled has a constant acceleration of 2.0 meter per second starting from rest.

a. How fast is it going after 5 seconds?
b. How far has it traveled after 5 seconds?
c. What is its average velocity in the first 5 seconds?
d. How far has it traveled by the time its velocity is 40 m/sec?

Objective 14: Value methods of graphical analysis by listing at least two reasons for using a graph to solve a motion problem.

We have said that graphs are often better for solving certain problems than using complicated equations. Below is a position versus time graph for three men, A, B, and C. List everything that this graph tells you about the men.

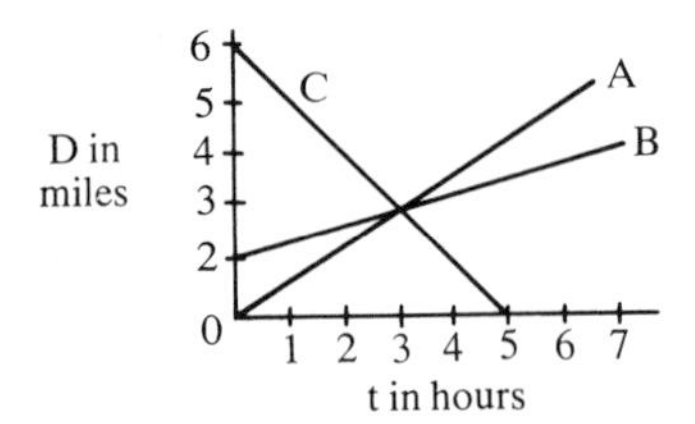

Objective 15: Develop confidence in the mathematical relationships which describe motion along a straight line by designing an experiment to confirm their reliability.

a. Describe the experiment you designed to validate the equation, $v_f - v_i = at$
b. Since no result is valid unless the experiment is repeated many times, which variables were changed and which variables were kept constant in your experimentation?

Objective 16: Develop confidence in his ability to use the measurings devices appropriate to this area of physics by stating the limits of error in his measurements.

Several instruments were used in the lab to study straight-line motion. Explain in detail how you used the bell timer. Be sure to include a discussion of how it was calibrated, what range of error you would expect, and how this error could be reduced.

Objective 17: Appreciate the role that generalization plays in scientific developments by listing at least two generalizations which apply to circular motion, drawn from the results of straight line motion.

An automobile moves with constant velocity around a circular track. The time to make one lap is T.

a. Write a mathematical expression for finding the distance the car traveled in a time, t_1.
b. How would you change this expression if the track were of the shape and dimensions shown below.

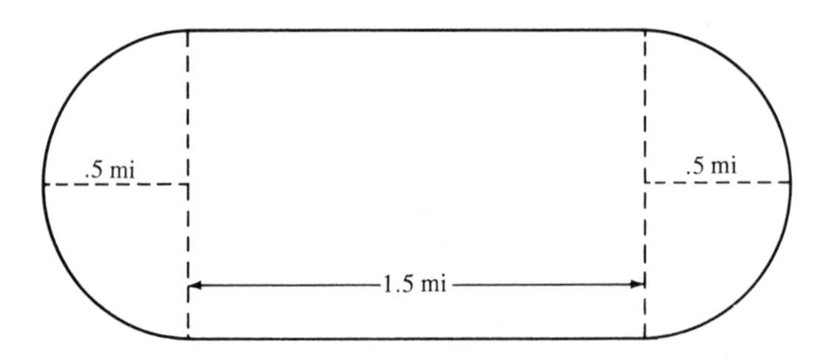

(1)

Chemistry

The following are examples of objectives and test questions for the Chemical Education Materials Study, chapter one and laboratory one.

Objective 1: The student should be able to accumulate information by writing clear, concise observations.

1. Which of the following is an observation about a lighted candle?
   a. The candle is made of carbon and hydrogen.
   b. Water and carbon dioxide are produced from a burning candle.
   c. The candle absorbed heat when it melted.
   d. The candle is made of a transparent white solid.
   e. None of the above.

Objective 2: The student should be able to differentiate between an observation and an interpretation.

2. Which of the following is an interpretation about a lighted candle?
   a. The candle gives off heat and light as it burns.
   b. The candle burns to produce carbon dioxide and water.
   c. The wick is made of three strands of string braided together.
   d. The burning candle makes little or no sound.
   e. The top of the candle becomes wet with a colorless liquid and the top becomes bowl-shaped.

Objective 3: The student should be able to recognize conditions that might affect the results of laboratory experiments.

3. Which of the following controllable conditions might influence the results on the observations of a burning candle?
   a. The experiment is done on the second floor.
   b. The lab bench is near the door.
   c. The windows are open.
   d. The room lights are on.
   e. The observation was made at night.

Objective 4: The student should be able to organize information and to observe regularities in it.

4. The following are the basic activities of science. A scientist performs all of them in a definite sequence. Put them in the proper order by placing the numbers 1, 2, 3, & 4 in the blanks.
   a. To communicate findings to others
   b. To organize information and seek regularities in it
   c. To accumulate information through observation
   d. To wonder why regularities exist

Objective 5: The student should be able to communicate scientific information qualitatively and quantitatively.

5. Read the five statements following. Place an "X" in the blank if it is a qualitative description and place an "O" if it is a quantitative description.
   a. The candle gives off heat and light as it burns._____
   b. The wick is made of three strands of string braided together._____
   c. The burning candle makes little or no sound._____
   d. The candle burns to produce carbon dioxide and water._____
   e. The top of the candle becomes wet with a colorless liquid and the top becomes bowl-shaped._____

Objective 6: The student should be able to carry on a process of inductive reasoning.

6. What is an elementary logical thought process framed on the basis of a collection of individual observations or facts?
   a. Inductive reasoning
   b. Deductive reasoning
   c. An experiment
   d. A generalization
   e. None of the above

Objective 7: The student should be able to organize information and observe regularities in it.

7. A lost child found out that tree limbs, broom handles, pencils, and chair legs all burn. He noted a regularity and proposed this statement. "Cylindrical objects burn." This statement could be called
   a. A theory
   b. A generalization
   c. Inductive reasoning
   d. Deductive reasoning
   e. None of the above

Objective 8: The student should be able to ascertain that there is uncertainty in all measurements.

8. Which of the following temperatures is probably the most accurate concerning the melting point of paradichlorobenzene?
   a. 53 degrees
   b. 53.2 degrees
   c. 53.203 degrees
   d. 53.2032 degrees
   e. None of the above

Objective for Questions 9–15: The student should be able to communicate scientific information qualitatively, quantitatively, graphically and mathematically.

Read: As the pressure on the piston in a tire pump becomes greater the volume of the air decreases.

Study the following four items:

A.

Pressure (in units called atmospheres)	Volume (in units called liters)	Pressure X Volume
0.100	224	22.4
0.200	109	21.8
0.400	60.0	24.0
0.600	35.7	21.4
0.800	27.7	22.2
1.00	22.4	22.4

B.

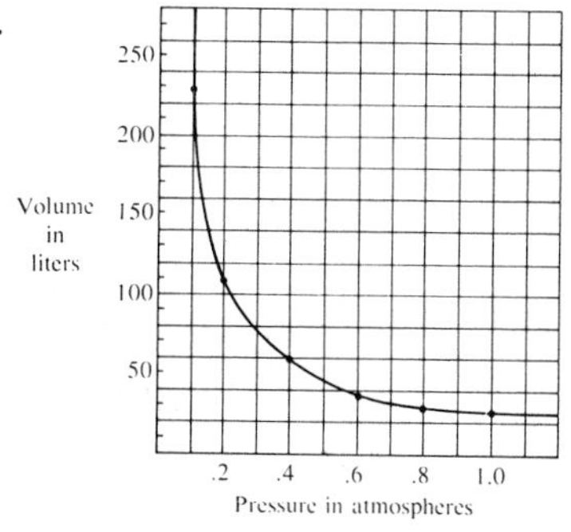

C. P x V equals 22.4 ± 0.6

P equals pressure in atmospheres

V equals volume in liters of 32.0 grams of oxygen gas at 0 degrees Centigrade

D. As the pressure rises the volume decreases.

9. Which of the above expresses the regularity between pressure and volume mathematically?
10. Which of the above expresses the regularity between pressure and volume quantitatively?
11. Which of the above expresses the regularity between pressure and volume graphically?
12. Which of the above expresses the regularity between pressure and volume qualitatively?
13. Which of the above is the least informative?
14. Which of the above gives the most detail about the pressure-volume behavior of oxygen gas?
15. Which of the above is probably the most useful?

Objective 16: The student should know that significant figures indicate uncertainty.

16. Which of the following numbers have four significant figures?
    a. 3609
    b. 3.609
    c. 0.3609
    d. 0.003609
    e. None of the above

Objective 17: The student should know that there is uncertainty in all four mathematical processes.

17. 12.3 is divided by 7.2. What is the most correct answer?
    a. 1.70
    b. 1.708
    c. 1.7083
    d. 1.70835
    e. None of the above

Objective 18: The student should be able to ascertain that there is uncertainty in all measurements.

18. Explain this term: 38.5 ± 0.2 degrees C.

Objective 19: The student should follow the rules of safety in the laboratory.

19. How should a thermometer or piece of glass tubing be inserted into a stopper? Explain why.

Objective 20: The student should be aware that even very simple appearing processes are very complicated and fascinating when subjected to careful observation and detailed description.

20. Go to the back table and get a container of cold water and place an ice cube in it. This seems to be quite a simple system; however, it may become quite complicated and fascinating if subjected to careful observation and detailed description.

    List *15 observations* that you can make from this ice-water system. Use the back of this paper.

Objective 21: The student should follow the rules of safety in the laboratory.

21. List five rules of safety to be followed in the laboratory.
    1.
    2.
    3.
    4.
    5.

(8)

Bibliography

1. Baker, Lyle. Unpublished paper, University of Northern Colorado, 1969.

2. Crow, Clem. Unpublished paper, Leeward Community College, Pearl City, Hawaii, 1970.

3. Duvall, Peggy. Unpublished paper, University of Northern Colorado, 1968.

4. Harszy, Arthur J. Unpublished paper, Ruby S. Thomas Elementary School, Clark County School District, Las Vegas, Nevada, 1967.

5. Lipson, J., Cohen H., and Glaser, R. *The Development of an Elementary School Mathematics Curriculum for Individualized Instruction.* University of Pittsburgh Learning Research and Development Center, 1966.

6. Phillips, J. and Zwozer, R. *Motion Geometry: Book 1, Slides, Flips, and Turns.* New York, Harper & Row, Publishers, 1969.

7. Picard, A. "An Analysis of the Objectives of a First Year Calculus Sequence, A Test for the Achievement of Those Objectives, and an Analysis of Results." Doctoral dissertation, The Ohio State University, 1967.

8. Shulene, John A. Unpublished paper, University of Northern Colorado, 1969.

9. "Summary Description of Grade Nine Science Objectives and Test Items," rev. ed. Edmonton, Alberta, The High School Entrance Examinations Board, Department of Education. March 1965, pp. 17-24.

10. Sund, Robert and Bybee, Roger. "Student Evaluation to accompany Earth Science Curriculum Project." Unpublished paper, 1969.

11. Winter, Stephen S. and Welch, Wayne. "Achievement Testing Program for Project Physics." *The Physics Teacher*, vol. 5, no. 5. May 1967, pp. 229-31.

12. Yamashita, June, Mathematics Department, Kailua High School, Kailua, Hawaii. Unpublished paper, 1969.

CHAPTER ELEVEN

EVALUATION OF OTHER EDUCATIONAL FACTORS

Evaluation of Teaching Skills

How do you recognize a *good* teacher? Some people observe characteristics of teacher performance, such as poise, appearance, vocal quality, etc. Others use student achievement as the criterion ("If the student hasn't learned, the teacher hasn't taught."). Johnson and Rising list five qualities necessary for successful teaching:

1. Competence in subject matter
2. Skill in communication
3. Dynamic personality
4. Acceptance and understanding of students
5. Competence in professional knowledge (4, p. 368)

While some of these can be measured with a "paper and pencil" test, numbers 3 and 4 require observation of the teacher in the classroom. Observations by a fellow teacher, the students, and the teacher himself shed light on different aspects of teacher performance and the classroom situation.

In the past, observation checklists centered on the externals of teacher performance, e.g., grooming, voice, etc. The renewed interest in behavioral objectives has prompted the development of observation checklists based on the cognitive content of the lesson. On the list below, the observer records instances of critical thinking and scientific processes. The observer places a check in the appropriate square each time the teacher requires students to respond in the manner indicated on the scale. For example, if the teacher asks a question requiring the students to apply knowledge, the observer places a check in the square after applying. The second time the teacher asks this type of question, the observer checks the second square. By following this procedure throughout the lesson, a profile of the type of questions asked and the kind of behavior elicited by the teacher may be obtained.

COGNITIVE EVALUATION FORM

Critical Thinking Processes

Applying						
Assuming (Asking children to recognize assumptions)						
Comparing						
Criticizing (analysis)						
Decision-making (analysis of what to do)						
Scientific Processes						
Classifying						
Collecting and organizing data						
Designing and investigation						
Formulating models						
Hypothesizing or predicting						
Inferring or making interpretations						
Measuring						
Observing						
Operational definition						

The evaluation form below is used in a manner similar to the previous checklist. Each time the teacher exhibits one of the behaviors listed, a checkmark is placed in one of the boxes.

PROSPECTIVE OR EXPERIENCED TEACHER EVALUATION FORM

The Teacher Should Be Able To:	1	2	3	4	5
1. *Demonstrate* enthusiasm in class presentations by speaking audibly, being well prepared, and zestful.					
2. *Display* tact and courtesy consistently.					
3. *Use* effective English.					
4. *Use* teaching skill that gets children involved.					
5. *Scan* the entire class during presentations and have good eye contact.					
6. *Face,* do not turn his back on students, when using the overhead projector or chalk board.					
7. *Maintain* discipline by having all children involved in educationally purposeful activity.					
8. *Give* positive reinforcement several times during class discussions.					
9. *Indicate* to students safety precautions when needed.					
10. *Ask* questions frequently requiring process type answers: i.e.: inference, hypothesis, etc.					
11. *Ask* more divergent than convergent questions.					
12. *Exhibit* a good background of subject matter by asking good inquiry types of questions.					
13. *Implement* and modify his teaching plans as needed by student responses so that they continue to be involved in class.					
14. *Move* close to students, is not behind demonstration desk, stands up during class discussions.					
15. *Ask introductory* questions or emphasize something the student should look for before films and filmstrips are shown.					

	1	2	3	4	5
16. *Ask* inquiry questions during and/or after audio-visual aids.					
17. *Demonstrate* fairness by applying class rules equally to all students and grading objectively.					
18. *Show* you his unit with teaching plans outlining behavioral objectives categorized according to Bloom's *Taxonomy* and activities to attain them.					
19. *Show* you measuring instruments for a unit indicating the evaluation of all the behavioral objectives including both cognitive and affective domains.					
20. *Show* you instructional materials he has created.					
21. *Demonstrate* that he can set-up and evaluate a laboratory experience.					
22. *Show* understanding of age group by adjusting questions so that children respond.					
23. *Cooperate* or team teach with other teachers and supervisors.					
24. *Spend* time after school preparing materials and laboratories for the next day or talking to students who wish to come in.					
25. *Give* evidence that he has spent time counseling students about their school work or futures.					

The following checklist lets the observer comment on a teacher's performance.

Mathematics Lesson Observation Sheet

Directions: The teacher characteristics in this form are grouped into four categories: 1) attention producing, 2) illustrating and using examples, 3) questioning and probing, and 4) reinforcement. Examples of each characteristic are given when appropriate. Comment on the presence or absence of each characteristic in the space provided.

1. Attention Producing
   a. Physical interaction with students: position of teacher in relation to students.
   b. Classroom mannerisms: verbal and facial expressions, gestures.
   c. Classroom interaction styles: type of participation required of the student.
   d. Pace and variety of lesson activities.
   e. Sensitivity to classroom behavior: teacher's awareness of inattention, boredom, or confusion.

2. Illustrating and Using Examples
   a. Sequencing of examples: from simple to complex.
   b. Relevance of examples to students past experience and knowledge.
   c. Appropriateness of examples: relevance to objectives of lesson.
   d. Use of examples to illustrate inductive reasoning.
   e. Use of examples to illustrate deductive reasoning.
   f. Use of counterexamples.

3. Questioning and Probing
   a. Level and number of questions: recall, higher level, evaluation.
   b. Sequencing of questions.
   c. Clarity and phrasing: length of question, precision of language.
   d. Adequacy of information and time for response.
   e. Handling of unexpected responses.
   f. Distribution of questions among students.
   g. Asking for more information on and/or meaning of initial remark.
   h. Asking for justification for a response.
   i. Refocusing attention on a related issue.
   j. Promoting interaction among students: asks second student to react to first student's response.

4. Reinforcement
   a. Positive verbal clues: words such as "good," "excellent," etc., using student's name.
   b. Positive gestures: smiling, eye contact, touching.
   c. Negative verbal clues: words such as "no," "wrong," etc.
   d. Negative gestures: frowning, showing impatience.
   e. Qualified reinforcers: handling an incorrect response without discouraging participation.

An interaction analysis form as shown below is similar to a cognitive checklist. This type of evaluation breaks down the interchange between teacher and student into observable behaviors. The lesson is then partitioned into timed intervals. The observer checks each of the behaviors observed in a particular interval. By investigating the analysis sheet for the entire lesson, the dominant types of classroom behaviors and the relative frequency of these behaviors can be determined.

INTERACTION ANALYSIS OBSERVATION SHEET

		Behavior	No.	1	2	3	4	5	6	7	8
CLASSROOM STRUCTURE	TEACHER	Lecturing	1								
		Asks Questions — (factual)	2								
		Gives Directions	3								
		Criticizes	4								
		Demonstrations	5								
		Explains	6								
	ACTIVITY	Accepts Judgements	7								
		Praises	8								
		Accepts or Uses Ideas	9								
		Asks Questions (Open Ended)	10								
		Guides Discussion	11								
		Supervises	12								
	PUPIL ACTIVITY	Response to Teacher (Relatively short—predictable)	13								
		Response to Teacher (Long and unpredictable)	14								
		Initiates talk to Teacher	15								
		Group Discussion with T-P interaction—Teacher not Dominant	16								
		Group Discussion with T-P interaction—Teacher Dominant	17								
		Teacher interacting with an individual	18								
	PUPIL ACTIVITY	Pupil initiates talk to another pupil	19								
		Response to another student	20								
		Group Discussion, No T-P Interaction	21								

		Lab		1	2	3	4	5	6	7	8
LABORATORY STRUCTURE	PUPILS IN GROUPS	Pupils working in groups—(no T-P interaction, some P-P interaction)	22								
		Pupils working in groups—(Some T-P interaction with the Teacher dominating)	23								
		Pupils working in groups (no interaction T-P or P-P)	24								
		Pupils working in groups (Some T-P interaction with the pupil dominating)	25								
	PUPILS INDIVIDU-ALLY	Pupils working individually (no T-P interaction, some P-P)	26								
		Pupils working individually (some T-P interaction T-P dominates)	27								
		Pupils working individually (some T-P interaction P. dominates)	28								
		Pupils working individually (no interest: T-P or P-P)	29								
		Confusion	30								
		Silence	31								
		Other	32								

(2)

Observations based on interaction analysis forms and cognitive checklists require practice. For best results:

1. Choose a partner by whom you can be observed without embarrassment and vice versa.
2. Discuss each item on the list and agree on its meaning; create examples of the behavior or situation.
3. Spend a few days observing each other without checklists to get the feel of the classroom.
4. Finally, make several observations over a period of time, say three weeks; use the "average" of these as the final evaluation.

Student evaluation of the teacher is often criticized because students generally lack the maturity and sophistication for accurate evaluation. The student, however, is the most frequent observer. His analysis sheds light on long-range teacher behavior not apparent to the trained, but infrequent, observer. Student evaluation also provides the teacher with feedback on understanding and acceptance of goals and objectives. A simple device for obtaining this kind of information is to give each student a copy of the course objectives and ask him to check those he feels he has achieved. By analyzing the responses of the entire class, the teacher can decide which objectives require further emphasis or alternate strategies.

Instrument to Determine the Effectiveness of the Teacher in Aiding Students Develop an Adequate Personality

Instructions: Tape record a class, lesson, or part of a school day while you are teaching. Listen to the tape & make notes under "Teacher development" about how you aided them in developing a particular area of their self-concept. Listen for teacher involvement. Rate these on a continuum — 5 for effective to 1 for ineffective.

Listen to the tape a second time and on form B note under "Observed behavior" the responses from students that behaviorally indicates they used the opportunity to develop their self-concept. Listen for student involvement. Again rate 5 for effective behavior and 1 for ineffective.

Put the two sheets together and compare your intended developmental procedures with the observable behaviors.

Idea Needed by Student	*Teacher Development*	Rating
1. Each person is trustworthy.		*1 2 3 4 5*
2. Children encounter situations where they succeed.		*1 2 3 4 5*
3. "Can-ness" is important.		*1 2 3 4 5*
4. New insights can develop from "mistakes"		*1 2 3 4 5*
5. Open communication for the child. (*Someone is listening.*)		*1 2 3 4 5*
6. Each child has an opportunity to make a contribution.		*1 2 3 4 5*

7. Each contribution is accepted.		*1 2 3 4 5*
8. There is a balance of common work and individual responsibility for specific tasks.		*1 2 3 4 5*
9. Children involved in planning. Student goals are considered.		*1 2 3 4 5*
10. Differences are desirable. Children volunteer new ideas.		*1 2 3 4 5*
11. NOW is important.		*1 2 3 4 5*
12. Desirable activity level. Children move about freely.		*1 2 3 4 5*
13. Evaluation is cooperative.		*1 2 3 4 5*
14. Evaluation is positive.		*1 2 3 4 5*

Idea Needed by Students	*Observed Behavior*	Rating
1. Each person is trustworthy.		*1 2 3 4 5*
2. Children encounter situations where they succeed.		*1 2 3 4 5*
3. "Can-ness" is important.		*1 2 3 4 5*
4. New insights can develop from "mistakes."		*1 2 3 4 5*
5. Open communication for the child. (*Someone is listening.*)		*1 2 3 4 5*
6. Each child has an opportunity to make a contribution.		*1 2 3 4 5*
7. Each contribution is accepted.		*1 2 3 4 5*
8. There is a balance of common work and individual responsibility for specified tasks.		*1 2 3 4 5*

Idea Needed by Students	*Observed Behavior*	Rating
9. Children involved in planning. Student goals are considered.		*1 2 3 4 5*
10. Differences are desirable. Children volunteer new ideas.		*1 2 3 4 5*
11. NOW is important.		*1 2 3 4 5*
12. Desirable activity level. Children move about freely.		*1 2 3 4 5*
13. Evaluation is cooperative.		*1 2 3 4 5*
14. Evaluation is positive.		*1 2 3 4 5*

(8)

Self-Evaluation Instruments for Science or Mathematics Teaching

Most teachers try to evaluate their own teaching, either formally or informally. If you've stated your objectives behaviorally, ask yourself if your teaching has helped the students achieve the objectives. Here are some specific questions to ask yourself:

1. Are the students aware of the course objectives?
2. Do my questions require only answers which could be memorized?
3. Do my students raise questions reluctantly?
4. Are the students aware of my criteria for assigning grades?
5. Do I teach my students how to learn?
6. Are my tests based on my objectives?
7. Are the students aware of what constitutes acceptable behavior in my classroom?
8. Do I encourage students to experiment with unconventional approaches?
9. Is the textbook the sole basis of my program?
10. Do I encourage students to seek information from many sources?
11. Are my examples chosen from non-textbook situations?
12. Does my personality or appearance hamper the student's understanding?

13. Do I treat my students with the same courtesy I expect of them?
14. Do I listen when students ask questions?
15. Are my objectives realistic for these students?

Evaluation of Tests

Teachers tend to rely on standardized tests for evaluating students. These tests claim to be highly reliable and valid and to have established national norms. The purpose of these norms is to allow the teacher to compare the student with other students his age and grade level. But tests do not always give an accurate picture of the changes which have taken place in the student. Traditionally, standardized tests in science and mathematics have measured the amount of information acquired by the student. For example, computation facts and skills in mathematics and facts about the physical world in science formed the bulk of these tests. Little or no attempt is made to measure process, attitude, the ability to conceptualize, creativity, or originality.

A teacher who uses commercially prepared tests uncritically is in danger of being dishonest with his students. Unless the teacher's objectives are identical with the ones used to construct the test, the test may not be valid for a particular class. The test must measure the behaviors emphasized by the teacher. To determine the behavioral emphasis of a test:

1. Construct a behavior-content grid based on the course objectives (see chapter three).
2. For each test item, consider:
   a. The student's content background and
   b. The amount of practice with a particular concept or process. (The cognitive difficulty of an item decreases with practice and familiarity).
3. Locate each test item on the grid.
4. Look for clusters of items or uneven distribution or untested teacher objectives.
5. Add or delete items as necessary.

The grid below was used to analyze a test constructed by the International Association for the Evaluation of Educational Achievement (IEA). The numbers in each cell refer to test items and the numbers in the category headings refer to the number of items in that category. A few test items follow the grid. Check their location on the grid. Would you have placed them differently? Why?

Behavior-Content Grid for Pre-University Students

	Computational processes	Verbal processes 31				
	Lower mental processees 41		Higher mental processes 28 Abilities			
	Computation 38	Knowledge 7				
Total 69	Skills and techniques in manipulation and computation 38	Knowledge and information about definitions 7	Translation and interpretation 5	Comprehension [follow and construct proofs, understand new developments] 4	Analysis 13	Application 6
New Maths 17	[22] [51] [30] [67] [35]	[41]	[43]	[33] [34]	[17] [31] [47] [25] [45] [52] [28] [46]	
Advanced Arithmetic 3	15 29 20					
Algebra 19	1 23 38 2 3 30 44 4 32 5 36 18 (37)		27	33 34	52	14 26 37
Geometry 5		12 13		8	7 21	

Analytical Geometry 9	10 61 67	55	11			
Calculus 9	57 63 58 64 59 65				62	56 60
Analysis 13	16 22 49 (19) 39 50 (40) 54 69	6	48	40	19 66	68
Trigonometry 3	42 53	9				
Sets 5	(35) 51		35 43		28 31	
Logic 6		24			17 45 25 46 47	
Probability 1		41				

Note: Items in square brackets [] appear in two content areas.
Items in round brackets ( ) appear in two process areas.

(5, p. 17)

Sample Test Items for IEA Test

Item 8

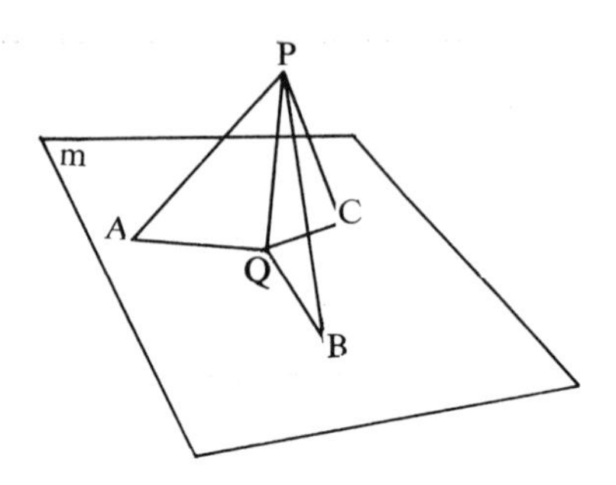

In the figure on the right, m represents a plane, and PQ is a straight line which is perpendicular to the plane at the point Q. Points A, B and C lie on the plane. If QA=QB=QC, then the triangles PQA, PQB, and PQC are

A. congruent (two sides and included angle)
B. congruent (two sides and angle not included)
C. congruent (two angles and corresponding side)
D. Similar but not congruent
E. Neither similar nor congruent

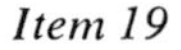

Item 19

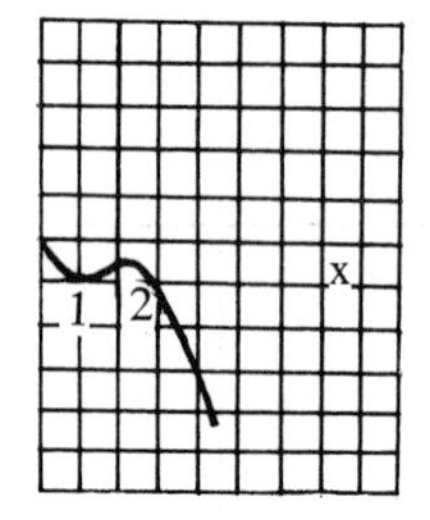

The graph on the right is the representation of one of the following equations. Which one does it represent?

A. $y = (1 - x)(x - 2)$
B. $y = (1 - x)(2 - x)$
C. $y = (1 - x)(2 - x)^2$
D. $y = (1 - x)^2(x - 2)$
E. $y = (1 - x)^2(2 - x)$

Item 27

A wholesale merchant bought a television set at a certain price and then sold it to a retail merchant at an increase of P per cent of this price. The retail merchant sold the set to a consumer for P per cent more than he paid for it. If the customer paid 65 per cent more than the price originally paid by the wholesale merchant, then P satisfies the equation:

A. $1 + \frac{2P}{100} = 1{\cdot}65$

B. $(1 + \frac{P}{100})^2 = 1{\cdot}65$

C. $1 + (\frac{P}{100})^2 = 1{\cdot}65$

D. $1 + P^2 = 1{\cdot}65$

E. $1 + 2P = 1{\cdot}65$

Item 56

An open cylindrical vessel of capacity 9000 c.c. is to be made with the curved surface of sheet metal and a wooden base. If the weight of 1 sq. cm. of the metal is three times the weight of 1 sq. cm. of the wood, each being of uniform small thickness, what will be the radius of the vessel (in cms.) when its total weight is a minimum?

Item 67

Find the difference b − a of the vectors a = (4, 2) and b = (0, 3).

A. (−4, −2) D. (4, 2)

B. (−4, 1)

C. (4, −1) E. (4, 5)

(5, p. 49–66)

Teacher made tests may also be analyzed by a behavior content grid. The following chart indicates how a teacher can use Bloom's *Taxonomy* to determine the sophistication of his testing procedures.

Although the chart on page 198 is designed to analyze the cognitive domain only, it can be easily modified to include the affective domain by adding a fifth category on the left hand side of the chart entitled: Affective Domain. Performing several analyses of tests using behavior content grids helps to change the perspective of experienced teachers involved in their utilization. It has been shown that the calibre of tests improves when constructed by teachers with this kind of experience (1, p. 37).

Evaluation of a Course by Students

Students may use the following instruments to evaluate various courses. An analysis of these instruments in conjunction with student self-evaluation inventories can aid the teacher in finding weaknesses in the instructional system.

The first and second evaluation forms were designed to evaluate elementary science methods courses and the third for an Earth Science Course. The fourth form asks the student to evaluate the instructor and the course. In addition, the student must complete a self-evaluation form.

1. Evaluation of the Elementary Science Methods Course

Directions: Please consider the contribution made by each of the following to your achievement of the objectives of this course. Mark an "X" on the scale where you would rate the effectiveness of each item. [Go to page 199.]

BLUEPRINT OF GRADE NINE SCIENCE EXAMINATION, 1965

OBJECTIVES	Topic or Content Area									Emphasis %
	Matter and Energy Force, Work & Power	Mechanics		Chemical Reactions	Heat	Light	Trans-porta-tion	Measure-ment*	Science* as Inquiry	
		Machines	Fluids							
1. *Knowledge*										40
2. *Comprehension*										30
3. *Application*										20
4, 5, and 6 — *Analysis, Synthesis and Evaluation*										10
Emphasis %	5	15	5	15	15	15	10	5	15	100

**Note:* The topics "Measurement" and "Science as Inquiry" cut across Content Areas. The latter category will be used for items that involve more than one content area or that involve inquiry as, for example, Items 6 to 10 under Analysis.

(7, p. 8)

1. The text, *Teaching Science Through Discovery*

	low			moderate			high	
1	2	3	4	5	6	7	8	9

2. Lecture-discussion sessions

1 2 3 4 5 6 7 8 9

3. Laboratory experiences

1 2 3 4 5 6 7 8 9

4. Laboratory supplements

1 2 3 4 5 6 7 8 9

5. Experimental nature of the course

1 2 3 4 5 6 7 8 9

6. Examination of new science curricula

1 2 3 4 5 6 7 8 9

7. Design of inquiry-oriented lesson plan

1 2 3 4 5 6 7 8 9

8. Construction of pictorial riddles

1 2 3 4 5 6 7 8 9

9. Writing of a unit segment

1 2 3 4 5 6 7 8 9

10. Construction of evaluation instruments

1 2 3 4 5 6 7 8 9

11. How would you rate this course in comparison to other methods courses you have taken?

1 2 3 4 5 6 7 8 9

12. How would you rate this course in comparison to other college courses in general you have taken?

1 2 3 4 5 6 7 8 9

Directions: Please write brief answers to the following questions:

13. In your opinion, what was the most important contribution made by this course to your ability to teach elementary science?

14. What was the least helpful part of this course?

15. If you were going to teach this course, how would you change it to make it better? (6)

2. Evaluation of the Elementary Methods Class

Directions: Please check the most appropriate column on the right to the following statements.

	Low	*Mod-erate*	*High*
1. The text, *Teaching Science Through Discovery,* contributed to my understanding of problems in teaching elementary science:			
2. The text, *Readings in Science Education for the Elementary School*, by Victor & Learner, contributed to my understanding of problems in teaching elementary science:			
3. The laboratory manual: *Elementary Science Teaching Activities,* I thought contributed to the course:			
4. The practice of having you teach science lessons to small groups of four or five students had the following value:			
5. The exposure to the Elementary Science Study: "Gases and Air" and "Mealworms" units had the following value:			
6. The exposure of the *Science Curriculum* Improvement Study Units had the following value:			
7. The exposure to the American Reading for the advancement of Science Process approach had the following value:			
8. The lecture time used for lectures, discussions, testing and showing films were of the following value:			
9. The time spent doing laboratory activities was generally of the following value:			

	Low	*Mod-erate*	*High*
10. The experience of writing a pictoral riddle was of the following value:			
11. The experience of writing a process-centered lab activity was of the following value:			
12. The experience of seeing how to use a film loop and a film in a discovery approach was of the following value:			
13. The experience of writing a unit was of the following value:			
14. The experience of using the evaluation scales in small groups was of the following value:			
15. The opportunity to rate yourself on the self-evaluation scale was of the following value:			
16. How would you rate this course in providing you with confidence and preparation to teach science:			

17. Write below any further comments you wish to make about the course:

18. If you were going to teach this course, how would you change it to make it better?

19. What is your over-all evaluation of the course?

3. Student Evaluation of the Earth Science Course

Directions: Please rate the following by marking the appropriate space.

SA — Strongly agree with the statement
A — Agree but not strongly
N — Neutral or undecided
D — Disagree but not strongly
SD — Strongly disagree

1. Too much emphasis is placed on topics that are unimportant in earth science — SA A N D SD
2. The problems in earth science are much too difficult — SA A N D SD

3. The textbook was really a great book	SA A N D SD
4. I am glad that I took this course	SA A N D SD
5. I like the amount of freedom I was allowed to have in this course	SA A N D SD
6. I am enthusiastic about earth science	SA A N D SD
7. I hoped that earth science would have been more exciting	SA A N D SD
8. I have learned a lot about how to work on my own and be more self-directive and responsible	SA A N D SD
9. This course has increased my interest in earth science	SA A N D SD
10. This course compared to others I have taken was excellent	SA A N D SD
11. This earth science class was dull, routine and monotonous	SA A N D SD
12. This class was better than other *science* classes I have taken	SA A N D SD
13. I am going to tell my friends to take this science class	SA A N D SD
14. I am generally more interested in science than before I took this class	SA A N D SD
15. Classroom discussions are more important than doing laboratory work on my own	SA A N D SD
16. The class activities were generally good	SA A N D SD
17. The laboratory investigations were interesting	SA A N D SD
18. The amount of time devoted to lab work should be decreased	SA A N D SD
19. The laboratory investigations were ineffective and contributed little or nothing to the class	SA A N D SD
20. This class had little or no opportunity for individual help or discussion from the teacher	SA A N D SD
21. Any time I needed help from the teacher, there was an opportunity to get the help I needed	SA A N D SD
22. I think the method used to teach this class was good	SA A N D SD
23. It was fun being confronted with a specific problem or investigation and then working on an answer	SA A N D SD
24. When the teacher lectured I learned the most	SA A N D SD

25. I really liked to discuss the topics in earth science	SA A N D SD
26. If the results you obtain in doing an experiment seem too far out, then they should be rejected	SA A N D SD
27. I didn't get to ask questions	SA A N D SD
28. I learned a lot about the scientific method	SA A N D SD
29. The scientific method really doesn't have much relevance to life outside the science lab	SA A N D SD
30. I think this science class showed me how important science is in my life	SA A N D SD
31. I am interested in learning more about science and scientists	SA A N D SD
32. The instructor in this class was extremely good	SA A N D SD
33. This science class was better than other classes because of the teacher	SA A N D SD
34. The materials made the class interesting.	SA A N D SD
35. The teacher had little or nothing to do with my high rating for the class.	SA A N D SD
There should have been more time in class for:	
36. Teacher lectures	SA A N D SD
37. Movies	SA A N D SD
38. Demonstrations	SA A N D SD
39. Laboratory investigations	SA A N D SD
40. Field trips	SA A N D SD
41. Tests	SA A N D SD
42. Reading	SA A N D SD
43. Discussions	SA A N D SD
44. Individual work in science	SA A N D SD

(2)

4. Colorado State College Rating Scale

TEACHER____________________________ COURSE____________________________

(NUMBER) (NAME)

DATE COURSE WAS TAKEN______________________________________

(SEMESTER/QUARTER) (YEAR)

To the Rater: You are asked to rate your instructor, the course, and yourself by responding to the items below. In order to indicate your rating, circle the

number for each of the points below which best describes your opinion according to the following representative scale:
(1) Inferior (3) Poor (5) Average (7) Above Average (3) Superior

I. *Instructor*

Item		Rating
1. Friendliness and pleasantness in manner	1.	1 2 3 4 5 6 7 8 9
2. Understanding and consideration for students	2.	1 2 3 4 5 6 7 8 9
3. Interest in development and achievement of students	3.	1 2 3 4 5 6 7 8 9
4. Interest and enthusiasm for course	4.	1 2 3 4 5 6 7 8 9
5. Knowledge of subject matter	5.	1 2 3 4 5 6 7 8 9
6. Daily planning and preparation	6.	1 2 3 4 5 6 7 8 9
7. Value of information presented by instructor	7.	1 2 3 4 5 6 7 8 9
8. Value of instructor's criticisms	8.	1 2 3 4 5 6 7 8 9
9. Promptness and regularity of instructor	9.	1 2 3 4 5 6 7 8 9
10. Ability to guide classroom discussion	10.	1 2 3 4 5 6 7 8 9
11. Ability to explain and lecture effectively	11.	1 2 3 4 5 6 7 8 9
12. Ability to interest and motivate students	12.	1 2 3 4 5 6 7 8 9
13. Adaptation of instruction to students' individual needs	13.	1 2 3 4 5 6 7 8 9
14. Fairness in testing and grading.	14.	1 2 3 4 5 6 7 8 9
15. Everything considered, including strengths and weaknesses, I would rate the instructor in this course as	15.	1 2 3 4 5 6 7 8 9

II. *Course*

Item		Rating
16. Organization of course	16.	1 2 3 4 5 6 7 8 9
17. Suitability of the text	17.	1 2 3 4 5 6 7 8 9
18. Suitability of the methods used	18.	1 2 3 4 5 6 7 8 9
19. Suitability of class size	19.	1 2 3 4 5 6 7 8 9
20. The extent to which course objectives were made clear	20.	1 2 3 4 5 6 7 8 9
21. The relationship betweeen objectives of the course and what was taught	21.	1 2 3 4 5 6 7 8 9
22. Student participation in determining course objectives	22.	1 2 3 4 5 6 7 8 9
23. Amount and type of assigned work	23.	1 2 3 4 5 6 7 8 9
24. How the course met my needs	24.	1 2 3 4 5 6 7 8 9
25. Opportunity for question and discussion	25.	1 2 3 4 5 6 7 8 9

III. *Myself*

Item		Rating
26. My opinion as to the worth of this course *before* taking it	26.	1 2 3 4 5 6 7 8 9

27. My opinion as to the worth of this course *after* taking it	27.	1 2 3 4 5 6 7 8 9
28. My study efforts for this course	28.	1 2 3 4 5 6 7 8 9
29. How well did previous course work prepare me for this course	29.	1 2 3 4 5 6 7 8 9
30. My approximate standing in the class	30.	1 2 3 4 5 6 7 8 9

(3)

Summary

Evaluation should not be limited to the achievement of objectives. Teaching skills, tests, and courses of study should be continuously evaluated, and modified, if necessary to improve the quality and relevance of instruction.

Lacking a specific definition of effective teaching, several aspects of teacher performance should be investigated. Attempts to evaluate teaching skills vary from studies of teacher performance, such as, grooming, voice, poise, etc. to studies of teacher-student interaction in the classroom. The latter type concerns itself with the kinds of questions asked of the students and the responses elicited by the teacher. In addition, the teacher's performance is studied for indications of the amount of student independence his teaching encourages.

Evaluation of tests is facilitated by a behavior-content grid. The grid enables a teacher to see which of his objectives the test measures. It also lets him see the emphasis placed on specific behaviors or content. Locating a test item on the grid requires that the teacher be familiar with the students background in content. The teacher must also know if the student has had little or much practice with the content. Items are generally placed lower in the taxonomy level as practice increases.

Teachers are often unfair to students when they use standardized tests to measure student achievement. The teacher must decide which of his objectives the standardized test measures. He must discard those portions of the test not related to his objectives. He must find a way to measure those objectives not covered by the test.

Students are intimately involved in course work. They may lack critical evaluation skills but they can provide valuable criticism of the learning situation. By combining the student's evaluation of the course and the student's self-evaluation on the course objectives, the teacher can obtain much information by which objectives might be changed to improve the course.

Bibliography

1. Baughman, G. D. and Mayrhofer, A. "Leadership Training Project: A Final Report." *Journal of Secondary Education*, vol. 40, no. 8. December 1965, pp. 369-72.

2. Bybee, Roger and Sund, Robert. Instrument for "Student Evaluation of the Earth Science Course," University of Northern Colorado.

3. Chalaupka, Donald. Instrument for "Colorado State College Rating Scale." University of Northern Colorado, 1970.

4. Johnson, D. and Rising, G. *Guidelines for Teaching Mathematics*. Belmont, California: Wadsworth Publishing Co., 1967.

5. Keeves, J. P. *Evaluation of Achievement in Mathematics*. Australian Council for Educational Research, no. 4 series no. 6, 1966. Permission to reproduce "Behavior-Content Grid for Pre-University Students" and "Sample Test Items for IEA Test" from International Association for the Evaluation of Educational Achievement, Stockholm, Sweden.

6. MacCormack, Alan. Instrument for "Evaluation of the Elementary Science Methods Course." University of Northern Colorado, 1969.

7. *Summary Description of Grade Nine Science Objectives and Test Items.* Rev. ed. The High School Entrance Examinations Board, Department of Education, Edmonton, Alberta, Canada, March 1965.

8. Stulp, Nina. University of Northern Colorado. Unpublished paper, 1967.

GENERAL INDEX

INDEX OF TESTING AND EVALUATION INSTRUMENTS